Rookie Year

Rookie Year

With a Foreword by
James M. Shannon
President, NFPA

National Fire Protection Association
Quincy, Massachusetts

Product Manager: Pam Powell
Developmental Editor: Robine Andrau
Editorial-Production Services: Paula Carroll
Composition: Publishers' Design and Production Services, Inc.
Cover Design: Cameron, Inc.
Manufacturing Manager: Ellen Glisker
Printer: Courier/Stoughton

NFPA No.: ROOKIE03
ISBN: 0-87765-482-4
Library of Congress Control No.: 2003106169

Printed in the United States of America
03 04 05 06 07 5 4 3 2 1

Rookie Year is dedicated to fire fighters everywhere.

Contents

Foreword

Through this book, you will come to know 12 men and women as they describe the training and experiences that made them fire fighters. These 12 people, who constitute a cross section of the estimated 1,000,000 members of the fire service in the United States, came from large cities, small towns, and very diverse backgrounds to join paid and volunteer fire departments.

Some of these men and women grew up wanting to be fire fighters. Others came to the fire service as EMTs, hospital orderlies, engineers, architectural students, teachers, photographers, foresters, and landscape workers. Regardless of where they started, the 12 fire fighters profiled in this book all share one adventure: the rookie year.

With candor and enthusiasm, the rookies tell how they decided to become fire fighters. They describe how they prepared for the physical rigors of fire training. They talk about their good days and their bad days. Throughout it all, their sense of service to the community is apparent.

The fire service is an important constituency for the National Fire Protection Association (NFPA). Because the work of the fire service is dangerous and complex, the fire fighter's ability to work effectively and safely is paramount to us. For that reason, *Rookie Year* includes three appendices: Career Opportunities for Fire Fighters, EMTs, and Paramedics, NFPA's *Standard for Fire Fighter Professional Qualifications,* and Fire Service Organizations in the United States.

Acknowledgments

NFPA is particularly grateful to those who told their stories and to their departments:

Cheyane T. Caldwell, Los Angeles City Fire Department, Los Angeles, California

Paul Christensen, Point O' Woods Volunteer Fire Department, Fire Island, New York

Andrew Couchman, Blue Township Volunteer Fire and Rescue Department, Pottawatomie County, Kansas

Pat Durland, United States Department of Interior, Bureau of Land Management, Boise, Idaho

Karen Estepp, Anne Arundel County Fire Department, Maryland

Jerry Horwedel, Los Angeles City Fire Department, Los Angeles, California

Jesus Moreno, Dallas Fire Department, Dallas, Texas

Arthur Moy, Cambridge Fire Department, Cambridge, Massachusetts

Joe Murabito, Delaware State Fire School, Dover, Delaware

Kyle Page, Carrollton Fire Department, Carrollton, Texas

Jennifer Steele, Kitsap County Fire District 7, Port Orchard, Washington

Dan Volcko, Phoenix Fire Department, Phoenix, Arizona

The writer who captured the rookies' stories so well is Clif Garboden. He is senior managing editor at the *Boston Phoenix* and brought the keen ear and skill of a beat reporter to his work on *Rookie Year*. He took a personal interest in this assignment, and the book is better for it.

NFPA is privileged to publish this book about the rookie experiences of 12 members of the fire service. I hope that you enjoy it and learn from it.

James M. Shannon
President, NFPA
Quincy, Massachusetts

Rookie Year

Two Rookie Years

1

Cheyane T. Caldwell

Cheyane T. Caldwell

PROFILE

Fire Department: Los Angeles City Fire Department

Location: Los Angeles, California

Position / Title: Fire fighter

Rookie Years: 2001–2002 and 2002–2003

Height / Weight: 6′1″ / 220 lb.

Previous Job: Fire fighter / paramedic

Hobbies / Other Interests: Weightlifting, waterskiing, camping, running, playing racquetball

Education / Training: B.A. in sociology, University of California, Los Angeles; M.A. in education, University of California, Los Angeles; James Sherman Fire Academy

Reason for Joining the Fire Service: "I decided that fire fighting was a more direct way [than secondary school administration] of helping people—a more immediate way of having an impact on their lives."

Most Memorable Rookie Experience: The "cookie factory" fire

Fire Service Ambition: "I aspire to become a Los Angeles city fire captain."

It was a big fire for a rookie—even for Cheyane Caldwell, a rookie who had more experience than most. Caldwell had just been out on a medical run, but his crew had been turned away from that and was headed back to the station when the truck was called to a fire in a nearby furniture factory.

Smoke loomed up over the fire site and Caldwell had no trouble spotting it from a good distance. When the truck arrived on the scene, it was raining, and smoke lay low to the ground all around the factory. The crew was told to run a 2½-inch hand line into the building. Caldwell took the nozzle and jogged toward the fire, only to discover that the front entrance was locked. While other fire fighters continued to work on the door, Caldwell and the line crew headed for the rear of the building. Eventually, he would drag more than 400 feet of hose to the fire, but even getting close wasn't easy.

"I came around the building," Caldwell recalls, "and there were flames coming around the corner down the back of the building from a large garage door 25 feet away. The fire was just pouring out and staying low because of the rain."

To make matters worse, there was a propane-fueled forklift inside the garage bay. Caldwell turned on his stream, but soon there was no water. The company had hooked the hose up to a dry hydrant. By the time an on-the-scene engineer had located the water valve under the street and turned the water supply to the hydrant back on, the fire had grown.

It took a long time to push the fire back into the building. Meanwhile, the forklift sat off to the side like a time bomb ticking. Until more of the fire was knocked down, the fire fighters could only spray the area around the machinery to keep it cool. Eventually, they were able to drive a front-loader into the still-burning building and get the propane equipment out.

The crews applied three 2½-inch streams and one 1¾-inch stream at the blaze and painstakingly brought it nearer to under control. As they worked closer and entered the building, roof tar melted and dripped onto their helmets. Low-hanging smoke superheated and burst into flame. Finally, a roof company was able to cut a hole in the roof for ventilation and another fire company broke through the building's front door.

Even then, nobody could venture more than 20 feet into the building. The fire fighters moved out and went to defensive streams in an attempt to contain the blaze from a safer distance.

Caldwell had been working the nozzle at the end of 400 feet of hose for 45 minutes and was exhausted. As the fire was coming under control, another fire fighter relieved Caldwell. The job, however, was still not finished. The crews spent several hours inside the building fighting spot fires. Rookies have had easier days.

Background

When Cheyane Caldwell entered the Los Angeles fire fighting academy in June of 2002, he had already completed one rookie year with the Phoenix, Arizona, Fire Department. Caldwell started out in Los Angeles, where he had grown up. He attended the University of California, Los Angeles where he majored in sociology and played the fullback position on the football team. After getting his undergraduate degree, Caldwell set out on two very different career paths. He enrolled in the master's program at UCLA, working on a degree in education. Simultaneously, he put himself through the James Sherman Fire Academy in Compton, California. By mid–2000, Caldwell had been prepped for both the classroom and the firehouse.

After what Caldwell describes as "a lot of prayer and a lot of talking with people," he applied for fire-fighting jobs in several cities.

"It's always been in my nature to go out of my way to help people I don't know," Caldwell explains. "I was thinking about going into secondary school administration, but I decided that fire fighting was a more direct way of helping people—a more immediate way of having an impact on their lives."

"I decided that fire fighting was a more direct way [than secondary school administration] of helping people—a more immediate way of having an impact on their lives."

"Besides," he continues, "I really like working outdoors. I worked on cars and enjoy doing things with my hands. I talked to a lot of people about the fire department and decided it was the place for me, a place where people had fun working together and had a blast fighting fires. In some ways, it's a thankless job. You show up at someone's house at 2 A.M. and help people who don't even know you. But it's a good feeling to make people's lives better—to help someone, even if you're strangers."

Phoenix Fire Department

Caldwell's first job offer with a fire service came from the city of Phoenix, Arizona, which has a department with a reputation for being progressive and for introducing a high number of modern innovations. In Phoenix, fire fighters are "empowered"—instead of doing everything by the book, they're encouraged to evaluate individual situations and tailor their activities to specific circumstances and hazards. All fire fighters, of course, draw a certain amount of distinction between what they learn at fire academies and how things are actually done in the field. The Phoenix Fire Department does have protocols and standard operating procedures, but it actively encourages its fire fighters to be creative and sees on-the-job problem solving as a learning and training resource.

Cheyane Caldwell had a mixed response to this unexpected approach. On the one hand, he loved the Phoenix department's atmosphere. "They treat their fire fighters like gold," he recalls. "The whole department really feels like a family. It's a wonderful place to work."

On the other hand, the focus on individual responsibility can make a beginner with a lot to learn feel insecure. Caldwell found himself seeking out the most demanding and experienced veterans on the force.

In Phoenix, recruits spend three months in the local fire-fighting academy, then move into active service for a probationary year that involves three firehouse assignments—one focusing on EMT responses, another with the emphasis on basic life support (BLS) training, and a third with a ladder company, practicing fire suppression.

Caldwell, an experienced student, did well at the academy. Upon his graduation, he was given some choice about where he would be assigned for his rookie rotations. "I asked for the harder houses," he says. "There were some houses that had reputations for having what I'd call 'sharks'—older guys who really make you look like a rookie. "Guys," Caldwell continues, half joking, "who ask you hard questions just to make themselves look smarter than they really are."

During his rookie assignments, Caldwell learned about firehouse pecking orders from bottom to top. "It's a standard relationship among the guys," he explains. "It's like on a football team, where the rookies carry the veterans' pads."

In many ways, Caldwell welcomed the experience and challenges of working within the firehouse hierarchy. For his truck rotation, he chose to go to a house with a captain who had a reputation for being the toughest in the city. "I really wanted to learn from that," Caldwell says.

"[The firehouse pecking order] is a standard relationship among the guys. It's like on a football team, where the rookies carry the veterans' pads.

"There was one guy," he remembers, "who was a 12-year veteran, a career fire fighter, who would jump all over you. He ran the house, and he picked on everybody. But he was a good guy really, and I wanted to challenge myself by working with the most experienced fire fighters."

Caldwell also wanted to have as much hands-on experience as possible. While he was attending the Phoenix academy, he took advantage of opportunities to go on house fire calls with the department. And when it came time for his probationary-year fire-fighting rotation, he asked to be assigned to Station 25, a firehouse in an area with a lot of single-family houses and a reputation for having a high incidence of fire calls.

The number of fires a station handles, however, is largely a matter of chance. While Caldwell was on probation in Phoenix, he saw very few fires, but he managed to learn from every opportunity that did arise.

On one occasion, Caldwell had just arrived for his shift, relieved another fire fighter, and tossed his gear on the truck when a call came in for a house fire. Caldwell's was the second truck to arrive and the crew was assigned to search and rescue. As they drove to the fire scene, Caldwell could see a looming column of smoke, indicating that the fire was already well established. "It was really cookin' when we got there," recalls Caldwell. When Caldwell's truck arrived at the scene, he quickly put on his gear, grabbed an ax, and headed into the burning house to search for trapped fire victims.

All that the rescue team found was a turtle, which they saved and delivered to the grateful children who had escaped from the house. The search was finished, but there was still some fire in a back room, which was being held in check by the first team to arrive at the scene. This was an ideal opportunity to break in a rookie, so the crew from Caldwell's station relieved the other fire fighters. Caldwell went in with the nozzle and came out with first-degree burns on his ears, a common fire fighter minor injury. They finished the fire, went back to the station, and critiqued the experience.

That was Caldwell's first serious encounter with a fire, and it turned out to be the last opportunity for him to confront a large fire in Phoenix. There were, however, lots of medical calls. The first one Caldwell responded to awakened him to the kinds of grim realities he would be confronting routinely as a fire fighter.

"It was the worst kind of call," Caldwell says. "It was an infant code—that means an infant death—and the first death I experienced on the job."

That call came in as an "infant cardiac arrest" in the local housing projects. When the medical emergency team with which Caldwell was riding got to the scene, they found a week-old infant lying on a kitchen table. The child wasn't breathing and had only a slight hint of a pulse.

"We scooped the infant up and took it to the ambulance. We did CPR all the way to the hospital, but it was too late."

Confronting tragedy is the expected burden of a fire fighter. Dealing with the emotional aftermath can be the hardest part. Was Caldwell very upset?

"Not really. It was a terrible thing, but our captain—who was a paramedic—was real good about it." Frequently, Caldwell explains, fire fighters use what outsiders might consider to be sick humor and dismissive attitudes to deflect emotional responses to tragedy.

"You learn right away that you can't take it home or you'll become a depressed person."

"You learn right away that you can't take it home," Caldwell says, "or you'll become a depressed person. Firemen are different animals. Not just anybody can handle dealing with all the crazy things we see. I had one captain who told us straight out: 'You have to let it roll off you like water off a duck's ass.'

"A lot of people don't get to that level where you're able to just live with it. Unless I really think about them, those things don't stick in my mind."

As most fire departments do these days, Phoenix offers fire fighters counseling, called *critical incident stress debriefing,* to help them deal with their emotional reactions to trauma and tragedy. Caldwell, like so many experienced fire fighters, says that—so far anyway—he's learned to "just deal with it."

When Caldwell finished his rookie year in Phoenix, he was assigned to be a "pool person," floating from station to station as needed. During his first months, the department decided to make an investment in him and funded Caldwell's paramedic school.

Los Angeles Fire Department

Caldwell was settling in and amassing experience and qualifications when he got a call from the Los Angeles Fire Department about an opening in the fire service back in California.

"It was the hardest decision," admits Caldwell. "I loved—I mean *really* loved—working for the Phoenix Fire Department. It was the best place I could imagine working. At the same time, I wanted to come home to LA where I could be near my family and my girlfriend. It was tough to leave Phoenix—really tough to leave such a good department with a progressive chief who wants everything to change with the times."

There was another major consideration. The L.A. department doesn't allow for automatic transfer in, so Caldwell, even with a private academy, the Phoenix academy, professional experience, and a paramedic certificate in his background, would have to start at the bottom of the ladder. He would have to go through the five-month training period at the Los Angeles fire academy and then experience another full rookie year, during which he would put in four months at each of three Los Angeles firehouses.

The pay-cut involved was minor. "Besides," Caldwell says, "money doesn't buy you happiness." Overall, however, there was no question that moving back to Los Angeles would be a career step backwards. "I wanted to come home," Caldwell says, "but I had to give up all that I'd gained and obtained." After a lot of soul-searching, Caldwell made the move to Los Angeles.

The flavor of the Los Angeles academy and the Los Angeles department was very different from what Caldwell had experienced in Arizona. In contrast to the sometimes experimental, learn-by-doing approach applied in Phoenix, the Los Angeles academy was run on an "old-school" military model. Caldwell didn't mind the change, even though it was noticeably different.

"I'd already had fire training, so a lot of what we covered at the academy was review for me," he says. "Still, I had to learn a new style. The academy was more like military training. LA also has what I'd call a more aggressive, and by-the-book style of fire fighting. There are more roof operations. The LA department prides itself on going inside. We go on the roof, cut a hole, and send fire fighters in to fight the fire."

Caldwell found the Los Angeles academy more like the private academy he had attended in Compton years before. Everything was strict; there was an emphasis on discipline and following standard operating procedures. Whereas the Phoenix department encouraged creative solutions, the Los Angeles department demanded its fire fighters to go by the book. "It's just a difference in style," explains Caldwell. "I'm comfortable with it either way."

Another difference, Caldwell discovered, was the role rookies played during their probationary year. In Los Angeles, he says, a rookie is supposed to "be seen, not heard. You're considered to be 'less' because you're new at the job. Even though I had experience, I had a lot to learn, so that made sense to me."

One thing Los Angeles did offer that Phoenix didn't was a large number of fire-fighting opportunities. "I've had tons of fire experience in LA," says Caldwell. "We had 14 apartments burning in one garden-apartment complex, a 100-by-300 square-foot commercial fire, a fire in a furniture warehouse, lots of house fires, and lots of arsons."

Fighting large fires, Caldwell learned first-hand, can be a physically taxing experience. At around 7:00 A.M. one morning, a call came in for a fire at a local cookie factory. Caldwell had come on shift at 5:30 A.M. and was busy cleaning equipment and doing a routine apparatus, making sure, he says, that "everything was high and tight."

From the station loudspeaker, a dispatcher rattled off a list of truck numbers. There would be many trucks at this fire. The task force (two engines and a truck) to which Caldwell was assigned was on the list. As they drove out onto the freeway, they could see the smoke above the fire site. As they approached, they were given instructions over the radio to go to a particular hydrant and hook up to it.

"That's a good thing about the L.A. department. They're aggressive and they expect you to be aggressive. They throw you right in and want you to have hands-on experience."

The factory was a tall concrete building with a 30-foot side. When Caldwell arrived, flames were already flying over the top of the building. They could look through an office door and see nothing but fire inside.

Following standard procedure, the crew threw a ladder against the building and extended their aerial ladder. Another task force had gotten to the roof first, so Caldwell and the crew he was with went up the other station's ladder to assist. The first step was to "sound" the roof to see if it was safe to walk on it. The fire fighters worked their way around the roof's perimeter until they got within talking distance of the other company.

By this time, fire was showing through the roof and the roof surface began to feel spongy under the fire fighters' feet—a sure sign that there was fire directly below. An apparatus operator started cutting a hole in the roof to ventilate the fire, a standard practice that gives smoke and fumes a path to escape so that fire fighters can enter the building more safely.

After a while, the apparatus operator handed the chainsaw to Caldwell, who took over the cutting. The saw jammed in Caldwell's hands and the chain broke. He had hit metal, probably a pipe or electrical conduit. Someone in the crew went off to retrieve another saw, but by this time flames were everywhere and the chief decided the roof was unsafe. It was time to "go defensive." The chief ordered all the fire fighters back to the ground and to set up a ladder-pipe operation.

A ladder-pipe setup allows fire fighters to direct a heavy stream of water at a fire from the top of an extended aerial ladder, giving fire fighters the advantage of working above or at roof level without having to risk being in or on the burning building. Once the ladder was in place, Caldwell's captain looked at Caldwell and said, "Okay, you're going up."

"That's a good thing about the L.A. department. They're aggressive and they expect you to be aggressive. They throw you right in and want you to have hands-on experience."

Fortunately for Caldwell, he had recently practiced the ladder-pipe procedure during one of the department's regular weekend training sessions. "The department's motto," he says, "is 'Train as if your life depends on it. Because it does.'"

"The department's motto is 'Train as if your life depends on it. Because it does.' "

Caldwell went up the ladder and hovered over the fire for more than 45 minutes, knocking it down with steady streams of water from above. The tactic worked and the fire came under control.

The chief called Caldwell down and the crew began cleaning up their equipment. The job wasn't over though. There were still spot fires inside the cookie factory, where stores of sugar and caramel had allowed the fire to become deep-seated. The burning sugars also created a lot of smoke. Caldwell and his fellow fire fighters donned breathing apparatus to enter the building to finish putting out the fire.

"My advice to anyone who wants to have a career in fire fighting is to take advantage of every chance you get to do the most you can. Be aggressive."

Four hours after he'd arrived at the fire scene, Caldwell was relieved.

It had been a big day. He'd been tested in a number of major live-fire situations—attempting ventilation on a treacherous roof, operating the nozzle from a ladder-pipe setup, and working small residual fires on the inside.

"Like I said," explains Caldwell, "LA is aggressive. They want you to have experience, so they had me do all that even though I was a rookie. I was the fifth man on the truck, but I was in the mix. They didn't keep me on the sidelines."

Caldwell had a similar feet-first experience on a medical call to a traffic accident on the freeway, where a victim was trapped inside a crumpled car. Caldwell grabbed the cutters and spreaders and attacked the sedan posts, literally cutting the roof off the car so the medics could extract the patient. "That was a senior job," he says. "But I just did it because I'd been trained and I knew the department wanted me to be aggressive. That's my advice to anyone who wants to have a career in fire fighting," he says. "Take advantage of every chance you get to do the most you can. Be aggressive."

Volunteer Spirit

2

Paul Christensen

Paul Christensen

Fire Department: Point O' Woods Volunteer Fire Department

Location: Fire Island, New York

Position / Title: Volunteer fire fighter

Rookie Year: 1997

Height / Weight: 6′2″ / 240 lb.

Full-time Job: Commercial photographer

Hobbies / Other Interests: Sailing, carpentry, auto racing

Education / Training: B.A. University of Colorado, Boulder, 1973; degree, Brooks Institute of Photography, 1975

Reason for Joining the Fire Service: "I always wanted to be a fireman."

Most Memorable Rookie Experience: Driving the truck full speed on the beach.

Fire Service Ambition: To help his community and learn new skills.

As a ski instructor in Colorado or as a commercial photographer in New York City, Paul Christensen never planned to be a volunteer fire fighter. But that was before he and his family began to spend their vacations on Fire Island, a 45-mile long barrier island running parallel to the southern shore of Long Island, New York.

Fire Island

At its widest, this sandy island between the Atlantic Ocean and Great South Bay measures about 1½ miles; at its narrowest, it's less than 50 yards wide. Motorized vehicles, except for emergency vehicles, are prohibited; residents get around on foot and by bicycle.

The "town" of Point O' Woods is small—¾-mile long and, at most, 150 yards wide. Located on the isolated eastern side of the island, the town comprises roughly 130 homes that are often no more than 20 feet apart. Homes are stick-built—balloon-construction summer homes that are neither winterized nor insulated. Often the homes have cedar shingles, long known to the fire service for their combustibility. The homes are built on pilings so they're 9½-feet above high tide, allowing a fire to draft air from below. Fires are rare, but when they occur they're devastating.

People who live in isolated communities – whether on an island or in a remote land-based area – depend on each other. "There's a strong spirit of volunteerism on Fire Island," Christensen explains. "There's no tax base and no central local government. We all have to contribute to the community—otherwise it won't work."

Each of the 15 communities and two incorporated villages on Fire Island must be on constant guard against both wildfires and house fires. Most of the Fire Island communities are private vacation villages and lack any official municipal status, so they've established small volunteer departments staffed largely by part-time residents.

Family and Background

Fifty-two-year-old Christensen is a successful self-employed professional commercial photographer, with a studio in New York City. He has been earning money taking pictures since he was fourteen years old and has a Bachelor of Fine Arts from the Brooks Institute of Photography in southern California. His clients have included Ralph Lauren, Huggies, Nickelodeon TV, practically every major pharmaceutical company, and documentary film producer Ken Burns.

"There's a strong spirit of volunteerism on Fire Island. There's no tax base and no central local government. We all have to contribute to the community—otherwise it won't work."

Eleven months of the year, Christensen lives in the upscale rural-suburban town of South Salem in Westchester County, some 60 miles north of Manhattan, with his wife and two children. Each July for the past six years, he and his family have lived in their modest vacation home in Point O' Woods. He returns to Fire Island most weekends throughout the year to work on the house, and while he's on Fire Island, he's on call with the Point O' Woods volunteers.

The first year that Christensen and his family summered there, he was invited to join the volunteer fire department. Christensen confesses that he harbored no lifelong dream of being a fire fighter. In fact, at early middle age, the invitation came as something of a surprise. "But I had been an Eagle Scout," Christensen notes. "I had had first-aid and lifesaving training. So it was a natural. Almost all new residents are asked to join. So like everybody, I had a personal desire to get involved."

Fire Training on the Island

Christensen submitted his application to the volunteer chief, and the next summer he found himself in training. Since Point O' Woods needs to maintain community insurance, all members of its volunteer department must become certified fire fighters. The fire-command center and training and research center for the district that includes Point O' Woods are on Long Island in the town of Yaphank. Instructors from Yaphank ride the ferry across Long Island Sound to bring classroom training to Fire Island volunteers, rotating classes among the island's communities.

"I had a personal desire to get involved."

"The instructors would show up toting overhead projectors and texts," Christensen recalls. Often the instructors would be retired career fire fighters from New York City who enjoyed recounting their experiences and entertaining the classes with awe-inspiring stories.

Christensen took his 40 hours of classroom training spread out over several "very long Saturdays" at the small Point O' Woods fire station. Just outside the station, other islanders on their way to the beach or the ferry waved at the students inside.

Despite the informality of the classes, the recruits took their training seriously. There were study groups within the departments where novices reviewed the class work and tested each other. "The study groups worked long and hard. We probably spent as much time in the study group as in class. We helped each other and rooted for each other at exam time."

"We probably spent as much time in the study group as in class."

At the end of the classroom sessions, a representative from the district fire command administered a written exam. When a rookie passed—with a minimum passing score equivalent to a C+—the Point O' Woods chief would order his turnout gear and he'd begin practical training. A self-described 200-pound 5-year-old, Christensen delighted in getting his own turnout gear.

During his practical training, Christensen remembers, the chief would always turn mistakes into an opportunity for putting fire-fighting tasks into the context of classroom training. What, he'd ask, would be the consequences if you did something out of order during a real fire? Then he'd have the rookie recite the rationale and philosophy for doing things by the book. Eventually, rookies went to the training center at Yaphank for their practical examination.

Together, the classroom and practical training took the better part of a year, during which the Point O' Woods rookies did not respond to alarms. The pace of training seemed slow, but, since the chief was overseeing part-time residents who had other jobs and obligations, it was practical.

Some of Christensen's practical training focused on fire-fighting techniques unique to oceanside fire situations. Christensen remembers being taught to defend against fires on the downwind side to prevent spreading—rotating in and out of the downwind position using portable extinguisher packs—because in dense seaside communities, a fire department's primary job is containment. Practical training, was, however, limited to exterior fire fighting, and no Point O' Woods fire fighter was expected to enter a burning building.

Point O' Woods volunteers repeat their training every year, starting with the basics, and everyone in the department is required to be recertified for CPR annually. Every Saturday morning, the chief holds some type of drill. In addition to practicing responses to residential fires and wildfires, Fire Island volunteers also train to fight fires in boats and to suppress fires on water by using foam.

Each fall, the departments on Fire Island cooperate in mutual-aid drills. The fire command at Yaphank always calls at least two departments for any alarm. Since alarms on the island involve fire fighters—though not usually equipment—from more than one community, mutual-aid drills are an essential part of the volunteers' training. Fire Island communities are private, and each department spec's its own gear. Consequently, there's no standardization of equipment from town to town. Hydrants and trucks vary widely and volunteers need to become familiar with them all so that one department can operate equipment in neighboring communities.

Point O' Woods Volunteer Fire Department

The political and social landscape of Fire Island adds dimensions of challenge to running a volunteer fire department. Politically, Point O' Woods has no civic structure as such, and the bulk of the population can best be described as "occasional." Membership in the Point O' Woods volunteer department numbers about 40, with at least two dozen members on hand on any given weekend.

The town hires a small staff to handle its maintenance and administrative chores; everyone hired is required to serve on the volunteer fire department. One of these town employees, currently a young carpenter, lives above the fire station, so it's continually staffed. Even the Point O' Woods fire chief is a nonresident. Like Christensen, the current chief lives in Westchester County. In the chief's frequent absence, one of the town's few paid civil servants, designated a "lieutenant," acts as chief. During the winter, there are always at least six employee-volunteers on hand. Since the smallest Point O' Wood truck has cab space for three, half a dozen volunteers are enough to roll.

As other Fire Island fire departments, the Point O' Woods volunteers respond to a lot of nuisance alarms because the high humidity and salt content in the air trigger automatic alarm systems. Since Christensen joined the Point O' Woods department, there has been one house fire a year; none, however, when he's been on call. In two of those instances, the structures burned all the way to the ground. Such an outcome is common in Fire Island communities, given the

"A house fire here is like a custom-made bonfire."

type of construction, closeness of the buildings, and the strong wind blowing off the ocean. "A house fire here is like a custom-made bonfire," says Christensen.

The department's most important job is to defeat the very real possibility that a single house fire could spread throughout the entire community. Much of the emphasis of Fire Island volunteer departments is on prevention. The highly combustible housing, combined with the dry dune grass, make Fire Island a real fire hazard. As a result, many communities on the island greet visitors with signs prohibiting grilling or even smoking outdoors.

Socially, Fire Island is a "party kind of place." A potential problem during the well-populated summer is that fire fighters will have been at a party before an alarm. The department, therefore, has strict rules about this situation: a volunteer who has been drinking alcohol is required to report to the station in any case, but is honor-bound to report his diminished abilities to the chief who routinely excuses him.

Special Challenges

The island's roads are narrow and primitive, designed for pedestrians and bicycles. Road conditions make the job of responding to fire that much more of a challenge. The island's topography and limited transportation system also make it a difficult environment for fire fighting. There is a service road running the length of Fire Island, but Christensen describes it as "little more than an unpaved alley." The fastest way to get from community to community is to drive on the sand above the high-tide mark. A normal fire truck—even if the volunteer departments could afford to maintain one—would be next to useless and would simply sink into the sand. Any long vehicle would be unable to navigate the tight turns between the houses.

Apparatus suitable for Fire Island has to be custom-ordered and designed to travel on unpaved surfaces. The vehicles carry water and have four-wheel drive and oversized, soft tires. They also have to be equipped to draft water from wells or the ocean. These vehicles also cannot be driven very fast on pavement—as can be observed when the truck goes to the mainland for servicing—but they can be driven on sand easier than ordinary trucks can. For use on most mutual-aid calls, the Point O' Woods department also has a GMC Suburban, that can carry eight fire fighters and minimal equipment when there's a greater need for staff than for equipment.

Commitment to the Island

The initial training, required recertification, and on-going drills are a big commitment for part-time residents to make. However, Christensen and other volunteers take the responsibility in stride.

"Point O' Woods isn't just a place for a quick vacation. Some families have been coming here for generations. Our friends are here, and our kids grow up here. Protecting this place and these people is important. And it's up to us," Christensen says.

Paul Christensen, having entered volunteer service in middle age, knows that his volunteer fire-fighting career most likely won't stretch beyond a decade. While there are no age limits set for Point O' Woods volunteers, participation is based on physical ability; the chief can ask for health recertification at any time. While he's able, Christensen finds volunteering fulfilling and fun.

"Protecting this place and these people is important. And it's up to us."

"I get to put on heavy black clothing and stand out in the sun. Believe it or not, I truly enjoy it. The exposure to procedures demystifies all the emergency work I see going on around me. That gives me confidence . . . and even exercise."

Christensen also has a "selfish" reason for loving his volunteer fire work. In the Point O' Woods department, if you're the first volunteer to respond and to put on your turnout gear, you drive the truck and operate the pump and the radio. Christensen's summer house is very close to the fire station, so he's frequently there first. "There's nothing like being in the center of everything," he says.

3

Volunteer Stepping-Stones

Andrew Couchman

Andrew Couchman

PROFILE

Fire Department: Blue Township Volunteer Fire and Rescue Department

Location: Pottawatomie County, Kansas

Position / Title: Fire fighter / EMT

Rookie Year: 2002

Height / Weight: 6′6″ / 210 lbs.

Previous Job: Student

Hobbies / Other Interests: Classic cars, fire memorabilia, playing basketball

Education / Training: Fire Fighter I certified, EMT training, Hazmat Awareness Level certified

Reason for Joining the Fire Service: "Prompted by the September 11 terrorist attacks, I chose fire fighting because it allowed me to protect my home directly and dovetailed naturally with my lifelong trajectory toward work helping others."

Most Memorable Rookie Experience: "Feeling dizzy when I was ventilating a roof during a structure fire and allowing a heavy metal crowbar to slip from my hands, skitter down the roof, slide through a hole, and just miss landing on my captain's head."

Fire Service Ambition: "To continue to serve the community whether in a paid fire service or volunteer service to protect life and property."

In his first year with the Blue Township Volunteer Fire and Rescue Department, Andrew Couchman went out on 150 out of 175 calls and routinely put in up to 30 hours a week at the station, cleaning trucks and maintaining gear between alarms. For this extensive dedication and duty he was paid . . . *nothing*. His officers did nominate him for the department's Rookie of the Year Award, which he received at the annual Christmas banquet, in December 2002, but Couchman's income for that year came from summer work with a landscape company and a job stocking shelves at Dillon Supermarket.

Like many volunteer fire fighters, Couchman made arrangements with his employers to shift his hours so that he would be available for daytime emergency calls during grass-fire season. He carries a pager with him constantly and no longer goes to bars because he wants to be completely sober if he's called to a fire or medical emergency.

Family and Background

Couchman left college—he'd been attending Kansas State University—to study fire fighting and to sign on with the Blue Township Volunteers. By the time he's completed a year and a half with the department, Couchman plans to have earned his Fire Fighter II and his Red Card certifications. The Fire Fighter II certification will qualify him for structure fire fighting and hazmat operations. The Red Card certification will qualify him to work for the government, fighting forest fires in the West. He is also enrolled in an EMT course.

Couchman has made this series of personal and professional sacrifices willingly. For him, the long hours of self-funded training and unpaid work are stepping-stones to a permanent position with a paid fire department. He hopes to gain such a position within two years of his signing on with the Blue Township Department.

Even with a paid fire-fighting job, Couchman knows that unless he moves to a major city on the East or West Coasts, he'll probably never make more than $40,000 a year. He and his fiancé, Rachael Ira, are already planning their life together accordingly.

"Fire fighting is something you do for the love of the job, not for the money."

"Fire fighting," explains Couchman, "is something you do for the love of the job, not the money," and Couchman has been gearing up for a life of public service since his infancy.

Couchman's first word, in fact, was not something simple and conventional like "Mama" or "Dada" or "ball." Instead, it was "ambulance." His father, now director of EMS in Riley County, Kansas, is a registered nurse and has been a certified paramedic for more than two decades. His mother has been a registered nurse for 24 years. His mother's father and uncle also were nurses. His fiancée, Rachael, whom he met at Kansas State, comes from a family well-populated with nurses as well. Couchman's little sister, Leah, is studying to be a paramedic and is already a certified EMT. For Andy, the oldest of the five Couchman children, health care and public service were expected career paths.

"I won't say there wasn't some pressure from the family," the 21-year-old Couchman says, "but, you know, when you're young you want to find things out for yourself. I went to college, but I was really attracted to fire fighting."

The terrorist attacks on New York and Washington on September 11, 2001, prompted Couchman to act. The realization that there was a real threat to personal security in his own country inspired Couchman to become involved. He saw two possible paths—military service or public service. He chose the latter because it allowed him to protect his home directly and dovetailed naturally with his lifelong trajectory toward work helping others.

Couchman talked with the chief at the paid fire department in his hometown of Manhattan, Kansas, about starting a fire-fighting career. The Manhattan chief advised Couchman to get as much experience as possible with a volunteer department that had a good reputation. The nearby Blue Township squad, the chief told him, was an exceptional training ground.

Blue Township Volunteer Fire and Rescue Department

"Like a lot of people, I always thought of volunteer fire departments as a bunch of redneck hillbillies driving around who didn't know what they were doing. And there *are* some volunteer departments that are like that. But not Blue Township."

The Blue Township volunteers, Couchman discovered, were well equipped, selective, and rigorous in their training requirements. The volunteer department has a strong sense of duty and responsibility to its community and a central role in the area's fire protection strategy. Even the fire stations are designed like firehouses for paid fire fighters.

Manhattan, Kansas, the state's original capital, with a population of 50,000, sits just north of Interstate 70 in Riley and Pottawatomie Counties, located roughly 50 miles west of Topeka in the northeast corner of the state. The Blue Township Volunteer Department has mutual-aid agreements with the Manhattan's paid department, eight or nine rural volunteer departments, and the part-paid / part-volunteer department in the city of Wamengo, for which Blue Township provides backup rescue support. Blue Township also aids 17 rural Riley County stations. On its own, the Blue Township Volunteer Fire Department covers a rural / suburban area measuring 50 square miles with a population of 4500. In addition to remote farm areas, the department answers calls in some of the fastest growing parts of the county as the suburbs extend out of Manhattan into the gently rolling Flint Hills. Not only does the Blue Township department respond to calls in all these regions, but it also often provides personnel to staff a paid firehouse while the station's crew is out on a call.

Having so many mutually dependent small departments scattered across a two-county area may seem inefficient, but the decentralization has actually proven to be an advantage, because it means that fire-fighting and medical-response resources are spread out to the point where some department is always available to respond quickly to a call anywhere in the region.

Couchman boasts that the Blue Township Department is a model volunteer fire-fighting organization that has won awards for its performance and organization. The department runs eight pieces of apparatus from two stations (one in the north of the township and one in the south). Blue Township's 28-member volunteer staff includes members age 18 to 65. Of them, 3 are fire fighter / first responders; 4 are certified fire fighters / EMTs; 2 are certified fire fighters and EMTIs (intermediate EMTs); 2 are fire fighter / paramedics; 4 are student EMTs; and 3 are currently prepping for EMT class. The remaining volunteers are strictly fire fighters with basic first-aid knowledge. By any volunteer department's standards, Blue Township's is a heavily qualified crew—a fact that's even more impressive when one considers that all that training ($650 for EMT certification; $6000 for paramedic training) was acquired at the volunteers' own expense.

Counting mutual-aid calls, the Blue Township volunteers respond to up to 200 calls a year, which comes to more than one call every other day. That's a lot of work for a volunteer brigade. The department's basic life support (BLS) unit is also on a first-response basis for Pottawatomie County. When a medical call comes in to the 911 dispatcher, the Blue Township medical volunteers are paged and on the scene—usually within 2 to 10 minutes—to provide emergency health care and pre-hospital care to patients. The Blue Township first responders don't transport these victims to hospitals, but the ability of the Blue Township medics to be on hand quickly—often before the primary medical crews arrive—is essential in the rural Midwest, where minutes are often, literally, a matter of life and death.

In the north of the township, Couchman explains, an emergency medical situation could easily be 15 minutes away from the nearest advanced life support (ALS) ambulance. But a Blue Township BLS squad, equipped with oxygen and an external defibrillator, could be on the scene within 10 minutes. Getting that equipment to a heart attack victim that much sooner could be crucial.

The Blue Township Department is also equipped to handle vehicle extractions. With State Highway 24 passing through the region, Blue Township is frequently called upon to apply its jaws of life, cutters, and spreaders to free victims from wrecked vehicles. They are also called to help with major extractions in the Manhattan area.

As with most modern fire departments, the majority of calls to the Blue Township volunteers are medical, but the squad gets more than its share of fire responses as well. The expanding suburbs keep both the Manhattan and Blue Township fire departments busy with common structure fires. Beyond the suburbs, the prairie and farmland—acres of dry grass susceptible to lightning storms and high winds—spattered with farmhouses great distances apart, create a challenging fire protection situation. As a city department Manhattan has only one truck large enough to fight a wildfire, so other departments, such as Blue Township's, are routinely involved in calls to nonhydrant areas.

Fighting a fire—whether a structure fire or a wildfire—in an area not protected by hydrants requires specialized equipment. The Blue Township Department has three engines; one holds 500 gallons of water, the other two carry 1000 gallons each. The key to working rural fires is the ability to shuttle water to and from the fire scene. For this, Blue Township uses a tanker that arrives at a fire with 1100 gallons of water and two collapsible reservoirs, which look like giant aboveground swimming pools.

Beyond the suburbs, the prairie and farmland—acres of dry grass susceptible to lightning storms and high winds—spattered with farmhouses great distances apart, create a challenging fire protection situation.

The tanker gets to the fire scene and deploys the reservoirs, fills them with water, and then heads back to a water supply for more. Couchman estimates that 80 percent of Blue Township's fire calls are to areas that are not protected by hydrants.

Maneuvering large trucks across prairie land is often difficult or impossible; so the department runs two "brush trucks." One is a heavy-duty pickup outfitted with a slide-in unit that holds 200 gallons of water preconnected to a run of garden hose. Although it lacks the volume capacity of standard fire hose, garden hose has the advantage of being lighter and smaller, so longer lengths can be carried by a small vehicle. Blue Township also uses a 2-ton 4-wheel-drive brush truck that carries 500 gallons of water.

The wildfire season begins early each year when Kansas farmers burn off the previous season's grass to make way for a new crop. For the farmers, this is a standard regional agricultural technique, but if the conditions aren't right, a set fire can go out of control quickly.

"You get some 80-year-old farmer who thinks he knows everything there is to know about farming and sets a field on fire when the humidity's low and there are stiff breezes and you've got a problem," says Couchman. He goes on to explain that to set a manageable field fire, the humidity should be high and the air should be still. Those conditions ensure that the fire will burn slowly and be easily contained.

Prairie grass in the area averages 3 feet high, which means that the flames of a prairie fire can range 3 to 15 feet. With a steady wind, such a fire can—literally—spread like wildfire.

A fire truck is lucky if it can move at 30 to 40 miles per hour over an open field. Under poor conditions, a grass fire can easily outrun even a light-duty fire-fighting vehicle. Prairie grass in the area, Couchman explains, averages 3 feet in height, which means that the flames of a prairie fire can range 3 to 15 feet. With a steady wind, such a fire can—literally—spread like wildfire.

On one such call, a farmer counted on the previous day's rainfall to slow his grass fire, but low humidity combined with 20- to 30-mile-per-hour winds took his field fire out of control. Fortunately, the farmer had a large tractor and was able to plow a 10-foot firebreak around his field. By the time the Blue Township volunteers arrived, their job was primarily to fight spot fires that jumped the fire line.

In another instance, a truck caught fire by the side of a rural road and exploded. Sparks set off a fire in an adjoining field that eventually spread over 2 miles and required two fire companies and 2000 gallons of water to extinguish.

Training

Because the Blue Township Department is so well equipped to protect its urban / suburban / rural area from such a wide variety of fire and medical emergencies, it seemed like the ideal proving ground for Couchman on his path to becoming a paid fire fighter. In early 2002, Couchman went to the first of two preliminary meetings with the Blue Township volunteers, where he was interviewed by the department's officers and fire fighters.

"At some volunteer departments," Couchman says, "you just sign up and that's that. But this department is competitive and selective. Some guys show up once and never come back."

"At some volunteer departments, you just sign up and that's that. But this department is competitive and selective. Some guys show up once and never come back."

His father had a solid reputation with the Blue Township squad, which helped at the first interview when Couchman explained why he wanted to join the fire service. At a second meeting, he was voted into a probationary position with the department and given a six-page checklist of tasks and procedures he had to master within 90 days.

During that time, it was up to Couchman to do the training when he could. Lacking any formal knowledge for mastering the tasks on that six-page list, Couchman relied on the help of Blue Township Assistant Chief Scott Emory, a 15-year volunteer veteran who joined the department after a 20-year Army career and splits his time between his fire-fighting responsibilities and a 50-hour-a-week job.

"I couldn't have done it without his help," says Couchman. "He always found time to work with me and help me to check off that list."

After three months, the Blue Township department's officers assessed Couchman's progress and granted him an additional 90-day probationary status during which he was invited to ride along on calls but prohibited from engaging in any frontline fire fighting. Above the rack where Couchman stored his gear at the Blue Township firehouse, his fellow fire fighters, following departmental tradition, hung a strip of ugly pink shag carpet.

"It looked like some pink carpet from the '70s," recalls Couchman. "Everybody could see it and everybody who came in asked about it. It meant I was a rookie, and everybody knew it."

During this time, Couchman withdrew from college and enrolled in a fire-fighting academy organized by several local departments. It was here that he passed the written and practical tests to earn his Fire Fighter I credentials. He also applied for positions with paid fire departments and discovered that he was on the right track because most departments required Fire Fighter I and EMT certification just to apply.

The Blue Township volunteers offer more training opportunities than do many other volunteer departments. The department holds three meetings a month. These are business sessions where volunteers discuss issues related to funding, staffing, and equipment, followed by several hours of rigorous training. The department maintains its own self-contained breathing apparatus (SCBA) training facility, which is a converted trailer donated by the community and elaborately outfitted as a three-level maze. For training, the department fills the trailer with smoke and designates one fire fighter (with breathing apparatus) to crawl in and play the role of "downed fire fighter." Volunteers, in full breathing gear, are then timed in their attempts to locate and rescue the victim. To add to the challenge, the passages in the trailer are rigged so an officer can change the configuration of the maze while the rescue attempt is in progress.

"Doing that, wearing 60 pounds of gear in 100-degree summer weather isn't easy," says Couchman. "After their first time, a lot of guys just turn in their equipment and say, 'This is not what I want to do.' Others want to do it over and over again. So sometimes we train late into the night—until 11:30 P.M. or midnight. Now that I'm experienced, I love to see the look in new guys' eyes when they suit up to do this for the first time."

To expand their training facilities, the Blue Township Department is currently building a concrete burn building, complete with propane-fueled fire simulation.

Rookie Year Experience

His rookie year included on-the-job experience as well. Couchman remembers his first call as a small, but important, learning experience. When he moved to the second stage of his probation, a Blue Township officer gave him a pager and explained that he was welcome to join in on any calls even though he wasn't yet qualified to do any serious fire-fighting work.

"I was just curious what the pager sounded like," recalls Couchman. "Then seven days after I joined, the thing beeped me awake on a Sunday morning in February. I made it onto the second truck out to a chimney fire. They let me ride in the back of an open truck. I had a roof over my head, but I was exposed to the weather. It was minus-20 degrees, and it had just snowed. I was frozen by the time I got to the fire."

"[Training in the three-level maze] wearing 60 pounds of gear in 100-degree summer weather isn't easy. After their first time, a lot of guys just turn in their equipment and say, 'This is not what I want to do.'"

What Couchman learned—and it was a valuable lesson, given the extreme low temperatures of Kansas winters—was that water can freeze in fire hoses and the water you spray can freeze on contact with the frigid ground. "Even though I couldn't really fight the fire, I learned a lot about slipping around," he says.

At the end of six months, Couchman had gotten a lot of training and became a full-time volunteer with the Blue Township Department. Completing the company tradition, the chief cut the pink carpet down from above Couchman's locker.

Once he was off probation, Couchman was able to put his fire-fighting training into practice. Still, he found, every call was a learning experience and on his most memorable day on the job, Couchman learned a lot. Things started off that day with a structure fire in a low-income residence. A spark from a washing machine had set clothes in a laundry room on fire, and the fire had spread to adjoining rooms.

Couchman, by then qualified for frontline work, was sent to the house's roof to ventilate the building. He stood there working his saw through the hot and buckling tar roof, praying that the joists were strong enough to support his six-foot-six, 210-pound frame plus the weight of his gear. The roof held, and the department was able to extinguish the fire in the area of the laundry room.

However, by this point the fire had spread to the attic, so Couchman stayed on the roof. As the volunteers were ventilating the upper story, Couchman was prying away roofing material with a four-foot haligan bar, a heavy metal crowbar designed for just such one-person demolition projects. Even without the heat from the fire, it was 95 degrees in the Kansas sun. Couchman felt dizzy. Before he realized what was happening, the bar slipped out of his hands, skittered down the roof, and slid through a hole. The tool just missing landing on the head of Couchman's captain.

The lesson—don't stay in a physically taxing environment too long before checking back to the truck for rehab—was reinforced with memorable humor when Couchman was later given the department's Bowling Pin Award. The award, a real 10-pin that hooks over the gear rack of the fire fighter who "wins" it, is handed out to a volunteer who messes up big time. The awardee has to keep the pin at his gear station and wear a bowling pin decal on his helmet until the next fire fighter pulls some equally boneheaded move.

"The chief is good about it, though. He says that if you don't get the award at least once, you're not participating enough."

The house fire lasted until about 4:00 P.M. Couchman and his fellow fire fighters returned to the station, cleaned their truck, scrubbed their hoses, and refilled their trucks with water. Exhausted, they were sitting around the soda machine relaxing when a second call came in—this one to a grass fire. By now the temperature was hovering around 100 degrees and the grass fire took an hour and a half to extinguish.

"I said, 'Okay, this is great. I love it!' but I was pretty tired," Couchman remembers.

"This isn't like a paid department where everybody works shifts. Because we're all volunteers, the same guys worked all three fires."

Andy Couchman's big day wasn't over, though. His pager went off again at 2:00 A.M. Lightning had struck a garage and set it on fire. "This isn't like a paid department where everybody works shifts," says Couchman. "Because we're all volunteers, the same guys worked all three fires."

The Kansas climate is a constant factor for fire fighters. Summer temperatures routinely run in triple digits and winter brings heavy snow and subzero cold. "It's tough to get out of bed in the middle of the night and go down in two feet of snow and scrape ice off your windshield to drive to the station," Couchman says. "But we do it."

There are some built-in compensation for all the hard work involved with volunteering with the Blue Township Department. Even the benefits, though, come in the form of hard work in rough weather. Each summer, Pottawatomie County is host to the Country Stampede, a country music festival by a lake at Tuttle Creek. For four days, the festival draws 50,000 people a day into the usually underpopulated state fishing area. Blue Township volunteer fire fighters provide around-the-clock emergency fire and medical protection for the event. In turn, they're given tickets to the show, which they usually barter to local farmers in exchange for the loan of all-terrain vehicles. The ATVs, loaded with medical supplies and hand-pump fire extinguishers, are ideal for moving through the crowds where trucks could never navigate safely.

"We camp out there for four days," says Couchman. "The festival is in June and it's usually around 120 degrees. Last year, we fielded 195 medical calls in four days. The combination of hot weather, lots of beer, and cowboys keeps us busy. Plus, there are a few trailer fires and the rowdy crowd building bonfires. It's actually the best-coordinated event held in Kansas."

The catch is that the Blue Township fire fighters have to use vacation time to work the music festival. According to Couchman, the volunteers are proud to do it. They even get to check out some of the performances while they're on break.

The amount of work—and most of it hard work—involved with working with the Blue Township volunteers surprised Couchman. "Some volunteers might show up only for calls, but it's all the work between calls that makes a volunteer department stand out," he explains. "At Blue Township, our trucks are clean. I'd eat off the floor of one of our trucks. You see some volunteer departments that show up in dirty trucks with soda cans rolling around inside and you know they don't have the kind of pride we do."

"Some volunteers might show up only for calls, but it's all the work between calls that makes a volunteer department stand out."

The real shock to Couchman came when he had to confront the emotions associated with dealing with people in trouble, especially people he knew, which in the close-knit area surrounding Manhattan, Kansas, happens frequently. On one highway accident call, Couchman approached a wrecked car to discover an old high school friend inside.

"That's when it struck me," he says. "This is real life. This isn't like *Backdraft* where you run into a burning building and save lots of lives. This isn't a movie."

Couchman's station is close to the area high school. Every time there's a call in that neighborhood, he worries that his brothers and sisters who are still in school might be involved. When an older Blue Township volunteer who'd been with the department for 25 years and no longer did any active fire fighting had a heart attack and drove his car into a tree in front of his home, it was the Blue Township medical team that came to his rescue.

"He was unconscious when the ambulance arrived and there's no doubt that the first responders saved his life," reports Couchman. "They could get a big head about that, but the crew takes it all in stride. In a lot of ways this is like a small town and you're going to deal with tragedies and people you know. You have to keep your distance—from good things and bad things. Sometimes the veteran volunteers seem brash, but they're not being insensitive—they're just protecting themselves from getting too close."

"In a lot of ways this is like a small town and you're going to deal with tragedies and people you know. You have to keep your distance—from good things and bad things."

Couchman has his eye on a paid firefighting career, but he's proud of being a volunteer and plans to continue his work with Blue Township even when he has a paid position in another department. Like many of his fellow volunteers, he takes public service very seriously.

"My friends say I'm crazy for all I give up," he says. "No drinking because I'm always worried my pager will go off, for example. They say I'm taking it too far, but the community depends on volunteers, and I feel the responsibility."

4

Career Jumps

Pat Durland

Pat Durland

PROFILE

Fire Department: United States Department of Interior, Bureau of Land Management

Location: National Interagency Fire Center, Boise, Idaho

Position / Title: Wildland fire fighter and smokejumper; currently wildland fire management specialist

Rookie Year: 1972 (fire fighter), 1975 (smokejumper)

Height / Weight: 6′ / 170 lbs.

Previous Jobs: Farm hand, survey crew, construction worker

Hobbies / Other Interests: Outdoor activities, snow skiing, biking, hiking

Education / Training: B.S. forest resource management, University of Idaho

Reason for Joining the Fire Service: "For the excitement, travel, scenery, and money (not necessarily in that order)."

Most Memorable Rookie Experience: "While enduring all the sweat and mental stress of training, the one thing that kept me going was the anticipation of making that first jump. When I thrust myself into the sky, all sound vanished and I found myself floating and tumbling in silence. It was truly a peacefully exhilarating experience! Before I knew it, I was on the ground but emotionally still high as a kite."

Fire Service Ambition: "I hope to affect bringing balance to our wildland fire policy and programs by proactively mitigating costs and losses before these events begin."

Pat Durland's fire-fighting career almost ended during his rookie year. It was 1972, and the federal wildland fire crew Durland had just joined was called into Fish Lake National Forest in Utah to put down a "timber fire," the technical term for a wildland fire involving acres of large trees. Durland was mopping up, a grueling job that involved knocking apart smoldering logs and mixing dirt with their embers.

It was a hot day, and Durland had made a rookie mistake—he had drunk all his water too soon and still had a long day ahead of him. Having grown up on a farm, Durland was used to hard work, but this was the hardest work he'd ever done, and he was hot—and very thirsty.

"I was miserable, and I really didn't know if this is what I wanted to do," Durland remembers. Just as he'd convinced himself that it was the end, that he would quit and find a less physically taxing job in forestry, a fire-fighting helicopter flew over, saw his plight, and dumped a load of water on Durland and the log he was working on. That changed everything.

"It was the best shower I'd ever had," says Durland, and his whole attitude changed for the better.

Today, Pat Durland is an internationally recognized expert in preventing unnecessary losses from wildland fires. He hasn't jumped out of a plane to a fire since the 1970s, but he's still fighting wildfires.

"Now I'm putting out fires before they start," the 52-year-old former smokejumper explains. "I'm using different tools—a phone and a computer instead of a chain saw. And instead of the weather, I deal with the political climate."

"Now I'm putting out fires before they start. I'm using different tools—a phone and a computer instead of a chain saw. And instead of the weather, I deal with the political climate."

Durland works for the U.S. Department of Interior's Bureau of Land Management (BLM), studying and educating governments, the public, and fire specialists about reducing fires started by people, the natural role of wildland fire, and how communities can exist compatibly in fire ecosystems. His work takes him to conferences and training sessions throughout the United States and the rest of the world. He's consulted in Canada, Honduras, and New Zealand, working with these countries to help protect communities and resources against unwanted wildland fires.

Durland's work involves working against the grain of traditional fire-fighting attitudes. "We've had this hero mentality where we wait for the event to occur and then respond by battling wildland fires," he says. "But to be effective, we must in the future shift from response to mitigation. We have to learn how to reduce losses before the event occurs—as we do with hurricanes, earthquakes, tornados, and other natural events."

For Durland, it's been a long road from throwing dirt in the western wilderness to being in charge of national program policy for a major land management agency. Along the way, he's had to invent his own career path and now finds himself on the front lines of a battle to change attitudes and behaviors toward wildland fires.

"To be effective, we must in the future shift from response to mitigation. We have to learn how to reduce losses before the event occurs—as we do with hurricanes, earthquakes, tornados, and other natural events."

Family and Background

Pat Durland grew up on the family farm in South Dakota. When he was a teenager, his family moved to the south-central part of Idaho. Durland's ambition was to live in Twin Falls, but the city, he discovered, didn't meet his expectations. Disappointed, he headed for the mountains, joining a survey crew with the Forest Service as a seasonal employee of the U.S. Department of Agriculture.

These survey crews laid out timber roads and trails through the remote Idaho mountains. Durland, who was studying civil engineering at the University of Idaho at Moscow, worked as a survey aid with an engineering crew. Everyone who worked in the field for the Forest Service went to basic "fire school," and everyone was expected to help fight local wildland fires. That's how 21-year-old Durland found himself deep in the wilderness, fighting forest fires.

The Salmon River, known to Native Americans as the "River of No Return," cuts east to west midway through the state of Idaho. It is surrounded by thousands of square miles of national and private forestland and the Salmon River Mountains, which, as Durland describes them, are "so steep you have to dig a hole to plant your butt and sit a spell."

Durland's first wildland fire experience came during his summer with the survey crew. He and 19 other forest service employees were called to a fire deep in the Salmon River Mountains. The 20 were flown to Cold Meadows airstrip on a DC-3, where they divided into two groups to board a smaller plane *Twin Otter* that would take them the last leg to the fire. It was dusk. As the small plane approached the fire camp area, Durland looked up through the cockpit window and saw what appeared to be an endless expanse of very steep mountains.

"I sensed that the pilot was getting serious," he recalls. "Then I spotted what looked like a little flat bar on the side of a mountain. That was our landing strip. I thought there was no way we could land safely on that short piece of flat ground."

Despite his concern, the plane landed safely, and Durland rejoined the rest of the crew on the ground. They were already cooking dinner. Durland had had only one week of fire training, and he was now working with an elite group of wildland fire fighters in some of the roughest country around. During the week he spent in the Salmon River Mountains, fighting that fire, he saw the real pros at work and was especially impressed by stories of smokejumpers.

The experience was inspiring. Durland remembers standing on a mountain top and looking down on planes as they dropped fire-retardant chemicals on the blaze below him. From afar he watched smokejumpers parachute into their landing zones and helicopters shuttle fire fighters from camp to the fire lines.

"Fire fighting was obviously more exciting than surveying roads and crunching numbers."

"It was energizing," he says. "Fire fighting was obviously more exciting than surveying roads and crunching numbers."

Durland switched his college major from civil engineering to forestry, a decision that served him well in the long run since the forestry curriculum required him to take natural and social sciences courses he would need later in his career.

"I never used my forestry education to count the trees as two-by-fours," he says, "but because of those science credits, I wasn't trapped in the technical series and was able to advance to a professional position."

After his taste of fighting fires, Durland asked to switch from engineering to fire fighting. Engineers on this forest crew, Durland was aware, weren't held in very high regard by the fire crew. "But the foreman concluded that this guy's judgment must be okay if he wanted to leave the engineers."

Training

The next season (1972), he was offered a job on the forest's fire crew. Durland's wildland fire experience intensified. The crew's base was in the mountains, and he was a given a list of gear he'd have to provide for himself. The list included boots, which Durland didn't have, so he went to Spokane, Washington, and got fitted for a 75-dollar pair of "Whites," the best you could buy. (In 1972, 75 dollars was a high price to pay for boots.) As the day Durland was to leave for camp approached, his boots had still not arrived. Without them, his feet would never survive the fire season. With the bus to Big Smokey base camp ready and waiting, Durland made one last dash to the post office. To his relief, his boots, which were to play a large part in his training experience, had arrived.

"Boots were very important," Durland explains. "The unwritten rule was to go out the first day and soak your boots by standing in a creek in the morning and then let the boots dry to your feet. And it worked. I never got blisters from those boots. They *were* the best!"

Durland and his fellow trainees spent two weeks in the classroom studying wildland fire-fighting techniques and safety. At the end of each day, they had physical training. This included calisthenics and running a mile to a nearby slope, trudging up and back down the hill, and then running back to camp, all with their boots on. "The guys who weren't in shape—and that was most of us—got back and just collapsed on the lawn and tried to keep our lunch down," Durland says.

After two weeks of classroom sessions, Durland and his new boots got a real work out when the training moved on to field exercises. The class would work "slash" fires and practice knocking them down and mopping them up. They also practiced what rookie wildland fire fighters do most—that is, building fire lines.

A major tactic in fighting wildland blazes is to contain the fire by clearing trees, grass, and brush in a swath along the edges of the fire. In most cases, it's impossible to get earth-moving equipment to these fires, so the work has to be done by hand, with shovels, pulaskis, and chain saws. Anyone who's ever tried to clear a bushy, rocky forest floor can appreciate the difficulty of building a trench for miles along the edge of a fire area. The job is exhausting and lacks mental stimulation, but it's a tried-and-true fire-fighting method. The hardest work invariably falls to the newest members of a wildland fire crew.

Sawtooth Crew

After training, Durland was finally a qualified member of one of two Inter-Regional Suppression (IRS) crews, the predecessors of what are now called Hot Shot crews. These nationally funded, regionally based teams of wildland fire specialists are dispatched to wherever they're needed across the United States.

The Sawtooth crew had a reputation for discipline. In the early 1970s, when men's hairstyles everywhere outside of Marine boot camps were long and shaggy, the Sawtooth IRS crew members were required to have short hair and were not allowed beards, mustaches, or sideburns. When they were together in public, Sawtooth crew members walked in straight lines. They looked sharp, but Durland remembers feeling out of place and conspicuous when the team entered restaurants in formation, wearing matching shirts and pants.

The second IRS crew, based in central Idaho, was the opposite, Durland recalls. "They all had long hair and moved in a mob," he says.

The protocols may have been uncertain, but these Forest Service crews represented a breakthrough in wildland fire management. They were the government's first modern attempt to organize specialist teams to fight wildland fires. Each crew had 20 members plus a crew foreman and assistant foreman and was divided into "squads." The Sawtooth crew hired some extra recruits to remain and provide fire protection in the forest when the crew was called to a fire away from the forest.

Gear was also basic. In those days, before nomex pants, Durland recalls having to supply his own Levis and sleeping "under the stars." Crew members carried 30-pound packs containing a change of clothes, a couple of "fireproof" shirts, a standard issue Army mummy-style sleeping bag, and a plastic groundcloth the fire fighters could pull over their heads if it ever rained. When the Forest Service first issued portable fire shelters to IRS crews, the Sawtooth crew used them as pillows. "They were hard, uncomfortable pillows," Durland recalls.

Crews learned the basics of fighting wilderness fires, which can be summed up in three procedures: throw dirt, cut brush, and dig fire lines.

Specialized crew training was also basic. The focus was on safety. The foreman offered case studies of fire fighters who had died in wildland fires. Many of these cases involved snags, or standing dead trees, which are famous for falling without making a sound until they hit the ground. Durland's foreman told his men, "If you die on my crew, it won't be because fire burned you, 'cause I won't put you in that situation. It'll be because you sat down to eat lunch under a snag."

"To this day, I look around for snags every time I stop in the woods," says Durland.

Crews learned the basics of fighting wilderness fires, which can be summed up in three procedures: throw dirt, cut brush, and dig fire lines. Sometimes they applied these procedures in the company of crews from other states. Along with the Sawtooth crew, the southern California crews, Durland recalls, had the best reputations for working fast and hard.

"When two hand crews met on the same fire," he says, "there was competition. It started with getting up the mountain to get on the fire line first."

Being on an IRS crew was a full-time commitment. On weekends, crew members were on "their own time," but still had to be available to respond to a fire. Durland recalls having to be "reachable" within a half hour at all times. These were the days before pagers and cell phones, so all emergency calls had to be dispatched by telephone landlines. Durland remembers being tracked down on the golf course by the clubhouse manager to respond to a fire. Announcements were relayed over everything from local radio stations to PA systems at drive-in movies.

"Your summer was not your own," Durland explains. Yet the crew members were glad to be interrupted on weekends. Weekdays, their hourly pay was fairly low, but after 40 hours on the job, crew members were paid time and a half. If they were fighting what was designated as an "uncontrolled fire," they were paid an extra 25 percent hazard duty differential. So weekend calls were considered bonuses, with crew members sometimes working 80- to 100-hour weeks, putting in 16- to 18-hour days.

The IRS crew pecking order had more to do with a member's fire-fighting experience than with his time on the crew.

Since the IRS crews were small, there wasn't a lot of room for drastic divisions of labor between veterans and rookies. According to Durland, the IRS crew pecking order had more to do with a member's fire-fighting experience than with his time on the crew, but Durland was new to both and started out at the end of the line.

"Rookies never drove the truck," he says. "Rookies also didn't get to do much sawing in the first year. First you had to prove yourself."

The basics mostly involved working with a shovel. Toward the end of his rookie year, Durland was given an unexpected opportunity to use a chain saw. "I started cutting everything in sight and thought, 'This is so cool,' " says Durland. "After about 45 minutes, I was beat and couldn't even grip the thing. It's like anything else—you have to learn to conserve energy, develop a technique, and pace yourself."

"After about 45 minutes [of cutting with the chain saw], I was beat and couldn't even grip the thing. It's like anything else—you have to learn to conserve energy, develop a technique, and pace yourself."

There was also a lot of work to be done just maintaining the IRS crew camp, such as taking care of the buildings, painting, and cooking. The rookies tended to get the brunt of the dirty work. Durland recalls, however, that those projects engendered camaraderie and pride in the operation, which ultimately brought both new and old crew members closer together.

That first year, in addition to the Utah fire during which the helicopter pilot determined Durland's career, the Sawtooth crew worked about a dozen major wildland fires. In the winter, Durland went back to school or did construction work.

In 1974, Durland's last year with the Sawtooth crew, he was called to work early in the year to join a crew in New Mexico, where the fire season starts earlier than it does in the Northwest. It was in New Mexico that Durland came into contact with a group of smokejumpers based in the town of McCall in central Idaho's Payette National Forest. The McCall jumpers followed the fire season around the nation, even supplying booster crews annually to Alaska.

McCall Smokejumpers

In the 1940s, smokejumpers were recruited primarily from the ranks of former specialized military paratroopers. By the 1970s, parachuting techniques and equipment had evolved to the point where average citizens were jumping out of airplanes for fun.

Durland had done some recreational skydiving and wasn't afraid of making it from the plane to the ground in one piece. He discovered that the Forest Service trained its own smokejumpers and signed people on, based on their forest fire-fighting experience, not their jumping skills.

"They can teach any goofball to jump out of a plane," he half-jokes. "What they want is people with training in fighting wildland fires. Smokejumpers are just regular wildland fire fighters who arrive at the scene by a different means."

"Smokejumpers are just regular wildland fire fighters who arrive at the scene by a different means."

So Durland set himself up for more training and another rookie year, this time with the McCall Smokejumpers. Unlike what he'd gone through with the Sawtooth crew, smokejumper training was more militaristic. Physical training for new recruits included running obstacle courses after work. At any time they could be told to stop what they were doing and race each other to suit up. The loser was rewarded with push-ups.

Training lasted about three weeks, and it was tough, especially the jump training. "When you jump out of an airplane," Durland explains, "you don't really have any sensation of falling because you're up in the air and nothing's close to you. But when you train, you jump off a 'shock' tower, and that's much scarier because you're close to the ground and there are all sorts of reference points around to give you perspective and the sensation of falling."

Shock-tower training came in the second week of instruction, along with other rough exercises. "They'd kick you out of the back of a pickup truck moving at 15 miles per hour," recalls Durland. "That would teach you to hit and roll."

"They'd kick you out of the back of a pickup truck moving at 15 miles per hour. That would teach you to hit and roll."

Recruits learned to maintain their jump gear, although in keeping with universal (and wise) paratrooper tradition, smokejumpers weren't allowed to pack their own 'chutes unless they survived their rookie year.

The jump training provided a dramatic contrast to the schooling provided at IRS crew camp, but not as dramatic as the behavior of the smokejumping recruits off the job. "I was used to the short hair, discipline, and team spirit of the Sawtooth crew, where we worked as a unit," says Durland. "These jumpers moved like an amoeba. Each was his own individual with his own opinion about how anything should be done, and it wasn't normal for them to think and work together."

On large wildland fires, smokejumpers are deployed to otherwise inaccessible areas of the fire's perimeter. Routinely, their missions are small group (two or more persons) assaults on small remote fires. Jumper crews patrol wilderness areas from the air, following thunderhead and lightning activity. They watch for small fires started by lightning strikes, usually high on the sides of mountains and far from trails, fire roads, and logging roads.

Durland's first live jump involved one of these two-person assignments. It was a classic smokejumper's experience and the kind of exploit thrill seekers dream about—straight from the pages of an adventure book.

As Durland and a veteran jumper prepared to jump, the "spotter" took Durland, who'd never jumped into a fire before, aside and told him to take care of his partner. It wasn't a joke or an attempt to distract Durland from being nervous. "Apparently, my partner," says Durland, "had a reputation for not thinking things through. So now the 'rookie' was unofficially in charge."

Durland's plane approached the small fire deep in the wilderness and, following standard procedure, dropped streamers to test the wind before the fire fighters jumped. The jump was made without any problems. Durland and his partner made their way to a large tree that had been struck by lightning and was burning fiercely at its top.

The process of dropping the tree and extinguishing the fire took all afternoon. Even small chain saws weigh about 40 pounds, and, because fire fighters have to carry out whatever gear they bring in, it's common to attack small fires with only a two-person crosscut saw.

Once the tree was down, the job of breaking up the log and extinguishing the fire with dirt began. As they worked, Durland reminded his partner to leave some fire burning so they could cook their dinner. True to his reputation, however, Durland's partner extinguished every glowing ember, so the two had to start over and build another fire in order to cook.

After eating, Durland and his partner shouldered their gear (about 100 pounds of jumping and fire-fighting equipment each), and walked to a nearby trail. They were met by a Forest Service packer with a team of mules and horses. The fire fighters packed their gear on the mules and rode the horses through 12 miles of wilderness to Chamberlain Basin airstrip, where they hopped a small plane back to McCall.

Changing Policies

This sort of suppression operation has traditionally been a staple of a fire crew and a smokejumper's work. However, Durland, who once earned his living fighting such small and remote fires, now advocates a more natural fire management policy that puts him at odds with the hundred year history of suppressing *all* fires.

According to Durland, the body of scientific and practical evidence suggests that natural fires actually benefit the many ecosystems by removing decadent fuels and stimulating new plant growth. If policy and practice require that all such fires be put out, growth continues unabated and the forest's fuel load increases. As the ecosystem tries to support more trees, the trees compete for water and nourishment. The forest exceeds its sustainability and becomes stressed. During drought, trees start to die and insects move in to do the natural growth-management job that fires would do more quickly and more cleanly.

"It's a dynamic process," explains Durland, who takes this message with him as he works with wildland fire management around the country. "Today, we're 99 percent effective in fighting wildland fires; but, as a result, when the 1 percent escapes our initial attack, the heavy, dry fuels burn hotter, are harder to control, and do more damage to these natural systems," Durland says. "More and smaller fires would be better than the large catastrophic fires we see these days."

"Today, we're 99 percent effective in fighting wildland fires; but, as a result, when the 1 percent escapes our initial attack, the heavy, dry fuels burn hotter, are harder to control, and do more damage to these natural systems."

Durland and other experts contend that we need both management and cultural changes in our wild ecosystems. They maintain we need to move from reactive suppression efforts toward progressive mitigations. Additionally, we need to concentrate on the things we can do before fires occur, such as establishing firewise communities and fuel treatment techniques.

"Idaho looks like Idaho because of wildland fire," Durland goes on. "just like Florida looks like Florida because of hurricanes. We have to learn to live compatibly with wildland fire in the West so it can continue to keep our systems healthy. We've done a tremendous job preventing structure fires, but in our fire-dependent ecosystems there should be no sprinkler systems in the woods because fire is part of the forest's natural process."

Part of the problem with such policy change is economic. For a wildland fire fighter, a "good" fire season (a year with a lot of fires) is a good year because it means a lot of overtime. Wildland fire fighters often find themselves in what Durland calls "overtime addiction," where they can't sustain their lifestyle without the extra money that weekend and 'round-the-clock work earns them.

"We have to learn to live compatibly with wildland fire. We've done a tremendous job preventing structure fires, but in our fire-dependent ecosystems there should be no sprinkler systems in the woods because fire is part of the forest's natural process."

Money isn't the only issue, of course. Most wildland fire fighters love their work. Durland describes the experience as fantastic, and explains that he was motivated by the thrill of the job. He loved the views he saw while flying over seemingly endless wilderness and jumping into the backcountry. Not only surviving but also doing the best job he could was tremendously fulfilling.

Wildland Fire Management

Durland left smokejumping after a few slow years, deciding that he couldn't throw dirt forever. (Although, he notes, in 2003 there were still jumpers at the McCall base who had jumped with him 28 years ago.)

In 1977, Durland went to work for the federal Bureau of Land Management at the Boise Interagency Fire Center (BIFC) now the National Interagency Fire Center, or NIFC. This job gave him year-round employment and the chance to use the knowledge and experience gained from a decade on the ground, fighting wildland fires.

After more classroom training at the Boise center, Durland became a regional aviation manager for BLM. He went from being a "groundpounder" to managing aircraft and developing national aviation and fire policy.

From there, Durland moved to a BLM post as an assistant state fire management officer (FMO) for California. In 1991, his career path led him back to Idaho, where the Bureau of Land Management was relocating national fire specialists, many from Washington D.C. The agency called Durland because of his expertise in wildland fire prevention.

Durland, who travels a lot in the course of his national responsibilities, is based in Boise, where he lives with his wife and two teenage daughters. As part of a federal agency, being isolated from the Washington bureaucracy is both good and bad, he says. It allows him the freedom to work without a lot of micromanagement, but it distances his department from potentially valuable political allies in Congress.

"We have to find a balance between wildland fire suppression and mitigation."

Still, Durland, who created his own career path in a fledgling area of the fire service, is grateful to be living in a relatively small city near the wildlands and is determined to apply his wildland fire-fighting experience to the mission of modernizing official policy and public understanding of wildland fire.

"Wildland fire agencies are lagging behind other disaster management organizations that take action to reduce losses before the event occurs," he says. "But it takes time to change the way people think and live. Our past 100 years of wildland fire suppression has won us many battles but is losing us the war for healthy forests, rangelands, and grasslands in the West," Durland adds. "We have to find a balance between wildland fire suppression and mitigation."

5

Overcoming Obstacles

Karen Estepp

Karen Estepp

PROFILE

Fire Department: Anne Arundel County Fire Department

Location: Maryland

Position / Title: Lieutenant, EMS technical response officer

Rookie Year: 1987

Previous Job: Student

Hobbies / Other Interests: Travel, history, museums, shopping, politics, being a mom

Education / Training: B.S., M.B.A., University of Maryland, CRT certification, Charles County Community College, EMT-P, Anne Arundel Community College

Reason for Joining the Fire Service: "I wanted to follow in my father's footsteps and do something exciting where I could make a difference."

Most Memorable Rookie Experience: "Realizing I didn't know everything and that I had a *lot* to learn and that book sense does not necessarily transfer itself into common sense."

Fire Service Ambition: "To be chief of my department, to promote change within the fire service, to change perceptions of people outside the fire service, and to know when I retire that I made a difference."

Estepp's first call came literally minutes after she'd been introduced to a mechanical CPR device—a tool she hadn't had while riding in Prince Georges County. "I had never seen nor heard of a mechanical thumper," Estepp says. "I was on my first shift, riding as the third paramedic to get acclimated to the system, and something told me to ask my partners how to set up and use the thumper. Ironically," Estepp continues, "after finishing the training and putting the equipment back together, we got a call for a cardiac arrest. I was so scared using the thumper that I was sweating profusely and my heart was pounding so fast I felt like I would pass out!"

Her first fire calls were equally unsettling. "I was scared. I knew I hadn't had much experience on the engine," she says, "and I really didn't want to look like an idiot. As I became acclimated, however, I began to feel more comfortable, although I still worried excessively about making a mistake."

Family and Background

Sometimes advantages turn out to be disadvantages. You'd expect Karen Estepp, daughter of M.H. Jim Estepp, the high-profile chief of Maryland's Prince Georges County Fire Department, to have an easy time joining the fire service. As it turned out, Estepp had to overcome two major obstacles: suspicions of favoritism and resistance to women becoming fire fighters.

Estepp had to overcome two major obstacles: suspicions of favoritism and resistance to women becoming fire fighters.

Jim Estepp became Prince Georges County chief at age 35 when Karen was 12 years old. He served as chief for 15 years, then as director of public safety for 2 years before returning to the chief's office. Despite his solid fire service background, even Chief Estepp was less than enthusiastic about his daughter becoming a fire fighter. He'd expected his sons might follow his career path, but Karen's two brothers weren't interested. Karen was.

For Karen Estepp, now in her late 30s and mother of five-year-old Scarlett, the rocky path to the fire service began with a University of Maryland first-aid course, offered as an elective in the school's criminal justice program. That training led to Estepp's riding along, as a third person, with a Prince Georges County EMT unit while she was still at the university and living at home.

She first set her sights on becoming a paramedic with the Prince Georges County service. To that end, she enrolled in a nine-month training program at Charles County Community College, where she earned cardiac rescue technician (CRT) certification. It was here that Karen Estepp ran into her first major obstacle. Prince Georges County has a nepotism rule that forbids immediate family members from working for each other.

Because her father was chief, Karen Estepp was denied a paramedic position. In an effort to remove the conflict of interest, Chief Estepp transferred hiring authority for paramedic positions to another office. Karen was turned down again, however. She appealed that decision, but was turned down for a third time at a personnel board hearing. During that appeal process, Karen Estepp got word that neighboring Anne Arundel County was hiring paramedics.

Prince Georges County has a nepotism rule that forbids immediate family members from working for each other.

Anne Arundel County

Anne Arundel County has a population of approximately 500,000. The county fire service, with 23 stations backed by 19 volunteer companies, serves the region's primarily residential areas. Anne Arundel County also includes several federal and independent departments.

As well as Annapolis City, Baltimore / Washington International Airport runs its own fire service. Fort Meade and the Naval Academy have federal fire fighters, but they dispatch calls through the county fire headquarters. With 500 miles of shoreline on the Chesapeake Bay, Anne Arundel County also maintains a marine division, complete with a dive-rescue team and fireboats. The county employs roughly 650 people, of which approximately 180 are paramedics and 15 to 30 are volunteer ones. The county borders the city of Baltimore and its densely populated southern sections are within easy reach of Washington, D.C.

Anne Arundel County maintains a big and busy fire service (in all, the staff numbers approximately 670 uniformed personnel), which responded to 70,000 calls in 2002—75 percent of them medical calls, dispatched from a centralized office at fire headquarters. When Anne Arundel County offered Karen Estepp a job in April 1987, she was hired as a lateral entry because the department desperately needed people with ALS credentials and Estepp had both training and volunteer experience.

Training and EMS Work

After a two-week orientation at headquarters, Estepp was assigned to a shift in the field doing EMS work. Shortly thereafter, she was assigned to rove throughout the county filling in for any station that was short staffed and pulling full 24-hours-on and 48-hours-off duty.

Despite this hands-on responsibility, Estepp was still required to enter fire fighter recruit school, and was enrolled in the first available class. Since the county department never had enough emergency medical personnel, she was assigned to EMS duty when she completed her fire fighter training.

Another career obstacle had to do with breaking out of EMS work to get experience in fire suppression.

At this point, Karen Estepp encountered two more career obstacles. The first had to do with breaking out of EMS work to get experience in fire suppression. The second involved being accepted as a woman in the fire service.

When she was first hired by the county, she was restricted to EMS work because she'd had advanced life support (ALS) training but no fire-fighting education. She wore a red stripe on her helmet, which signified that she could work in the field but couldn't go near fires. However, because fire suppression training is required for everyone in the Anne Arundel County Fire Department, Estepp, as a recruit, got the same exposure to fire fighting as did the other recruits. During her six-month academy experience, she lifted the same ladders and met the same demanding physical standards as did the men in her class.

Yet, even after that training, the department was reluctant to take Estepp off EMS service. At that time, even EMS officers weren't asked to fill in when a fire officer was absent. Fire officer duties were instead assigned to one of the fire fighters. That policy was eventually changed, but as a rookie Estepp felt hemmed in by the EMS career path. Although she continued to work with EMS services and became a paramedic in 1991, after the department offered to pay for her training at a community college, she has since gone on fire calls and acted as a fire suppression officer.

Gender Bias and Perceived Favoritism

The larger obstacle Estepp faced was gender bias. In 1987, female fire fighters were rarer than they are today. Joining a primarily male department was more of an uphill struggle than Estepp had anticipated.

In 1987, female fire fighters were rarer than they are today.

During her academy training, Estepp confronted, and overcame, her own mental obstacles—doubting her ability to throw a 35-pound ladder and to keep up physically with her male classmates. But, Estepp recalls, the social part was the hardest. "I was told, 'If people don't pick on you, they don't like you.' Most of the jokes and teasing were in fun. But some people were malicious, and I knew I was a target just because I was female."

"Most of the jokes and teasing were in fun. But some people were malicious, and I knew I was a target just because I was female."

Things were even harder for Estepp back at the fire station, where she was the only rookie in the house. In addition to being the target of traditional rookie-hazing pranks such as shortsheeting, Estepp felt she was facing something deeper and more serious. "I felt out of place," she says. "Some people—especially some of the older veterans—made it well known they didn't want women around."

At the time, there were only ten women in uniformed service in Anne Arundel County. Women slept in dorm rooms with the male fire fighters. There were no separate washroom or shower facilities for women. At some stations, there were no locks on the bathroom doors. Women were expected to put a sign reading FEMALE on the door when they went in. Some of Estepp's fellow fire fighters would challenge her with politically incorrect comments, as if, she remembers, they were saying, "I dare you to do anything about it."

"When you're new and have some obstacles to overcome, the last thing you want to do is make waves," Estepp explains. "I think some people thought I'd quit. And I do remember asking myself if maybe I'd made a mistake."

Estepp's problems with being accepted as a female rookie were aggravated by the common knowledge that her father was an important

"Some people—especially some of the older veterans—made it well known they didn't want women around."

and influential figure in the neighboring county's fire service and government. When she was a medical volunteer back in Prince Georges County, some of her co-workers there thought she was a spy for her father. Her family reputation followed her to the Anne Arundel County Fire Department, where she was teased (and not always affectionately) that she'd gotten her job because of her father's influence—ironic, of course, because Estepp had actually been denied a job in her home county because of her family connection.

Against this difficult social backdrop, Estepp began to question herself and came to believe she didn't have a lot in common with the people she was working with. She made social mistakes. Estepp had come from the busier Prince Georges County department, where staff with CRT certification could administer more drugs. She'd had field experience with gunshot wounds. Mentioning such things on the job in Anne Arundel County only made Estepp unpopular.

Work Experience as a Rookie and Beyond

"I had more education than experience," Estepp recalls. "I'd read a lot and traveled a lot. I felt isolated."

"I had more education than experience."

Although Estepp eventually committed to a five-year contract working with the EMS in exchange for the county's funding her paramedic training, she's since learned her way around a fire scene and has served as an officer for at least one three-alarm fire.

Slowly, Estepp gained experience and confidence on the job. Times also have changed. Today, she reports, her department has made great progress in terms of accepting women. The knee-jerk prejudice against female fire fighters is less common among younger members of the service. It was once assumed—usually incorrectly—that physical standards at the academy were lowered for women. Today, everyone knows first-hand that all recruits, regardless of gender, must meet the same fitness and agility standards.

All the time Karen Estepp was struggling with job-related challenges and occasional insecurities, she never lost confidence in herself.

"I stayed with EMS, and I love the work, but I get bored doing the same thing for a long time. And I don't really work well in groups," she confesses. "I have a wide range of interests, and I like to work alone. I like to be in charge, set my own deadlines, and do things my own way."

It was once assumed—usually incorrectly—that physical standards at the academy were lowered for women. Today, everyone knows first-hand that all recruits, regardless of gender, must meet the same fitness and agility standards.

Over the years, Estepp's admitted loner tendencies have paid off. Starting with work under her first EMS division chief, Roger Simonds, a career mentor and teacher during her paramedic training, Karen Estepp has taken on a series of special projects through the county fire department. She's helped develop quality assurance software used by the International Association of Fire Fighters, done paramedic training, and worked on developing EMS accreditation programs. Carrying the title of Lieutenant Advanced Life Support Training Coordinator, she has also put together the department's current ALS program.

Estepp's special projects work follows her wherever else her career may go. After the terrorist attacks of September 11, 2001, Estepp's chief was appointed to a four-year position on the Maryland Security Council, which assesses the state's ability to handle terrorist attacks and related emergencies. Karen Estepp moved with him as his assistant, but she continues to perform her EMS duties, which consist of being involved in the Anne Arundel County Fire Department's quality assurance and EMS certification programs.

"I still do everything from my old job that not many people have been trained to do," she says, not complaining. With a young daughter to raise, she concedes that a heavy workload—even if it's day work—is an additional challenge. But Estepp, who was doing shift work when Scarlett was born, isn't worried. To her, the hurdles look lower all the time.

In January 2003 the department underwent a reorganization, resulting in the creation of six new positions with the title of EMS technical response officer. Estepp was one of those selected. "It has provided a welcome change, new challenges, and new opportunities to effect change," Estepp says. "And it also allows me to be back on shift work."

Rigors of Training

Jerry Horwedel

Jerry Horwedel

PROFILE

Fire Department: Los Angeles City Fire Department

Location: Los Angeles, California

Position / Title: Fire fighter

Rookie Year: 2001–2002

Height / Weight: 5′8″ / 165 lbs.

Previous Jobs: College student, hand crew

Hobbies / Other Interests: Dirt bike riding, mountain biking, water skiing, golf

Education / Training: Mechanical engineering at California Polytechnic, San Luis Obispo, California; Los Angeles Fire Department Academy

Reason for Joining the Fire Service: "I was born into the fire service. It was the only thing I wanted to do since I was old enough to understand what my father did for a living."

Most Memorable Rookie Experience: "First day at Fire Station 26."

Fire Service Ambition: "To serve the citizens of Los Angeles to the best of my ability."

The first time Jerry Horwedel entered a burning building, he tripped over a couch. The call came during the second rotation of Horwedel's rookie year. He was training at Los Angeles Station 91, which runs only one engine and an ambulance, and was assigned to be the "nozzle man" on the call. As the engine Horwedel was riding approached the fire scene, he could see massive clouds of smoke billowing into the sky. Station 91's engine was the first to arrive. Horwedel grabbed his hose from the engine, which carried 500 gallons of water, and headed for the front door. It was locked. Another fire fighter yelled that the back door was open, so Horwedel repositioned himself and headed in.

"It was completely dark and full of smoke," Horwedel recalls. "I couldn't see a thing. I had no experience, and I fumbled through the room. I crawled over the sofa and tables looking for the fire."

Before long, a ladder truck from another station arrived at the scene, and a crew cut a hole in the roof to let the smoke escape.

"The smoke cleared like magic and I could find the fire," the 24-year-old Horwedel says. "When I looked back at the route I'd taken through the house and how many obstacles I'd walked into, it was almost funny."

Family and Background

"I was born into the fire service," Jerry Horwedel says. His father, uncle, and grandfather had been Los Angeles city fire fighters. Two uncles were Los Angeles county fire fighters. Jerry was expected to grow up to join the fire service. "It was the only thing I wanted to do since I was old enough to understand what my father did for a living."

As Horwedel grew up in Lancaster, north of Los Angeles, he never even considered another career path. However, the Los Angeles City Fire Department is difficult to join. It was Horwedel's father who suggested he have a backup plan, which is how Jerry Horwedel became the first member of his family to go to college.

"I didn't even know what mechanical engineering was," he says. "but one of my teachers said I was good at math and should look into it."

At California Polytechnic, in San Luis Obispo, Horwedel found that he loved mechanical engineering. Suddenly, he had an unexpected career option. He also met his future wife, Stephanie, at school. She had no family or other connection with fire fighting and feared the dangers of the job. Horwedel was torn—should he follow his lifelong goal into the fire service or concentrate on his mechanical engineering degree?

In the end, it was no contest. As much as Horwedel enjoyed engineering, when the opportunity to attend the Los Angeles City Fire Department Academy came, in the last semester of his senior year, Horwedel dropped his studies, one course short of his degree. Although he can—and likely will—finish his education at some point, putting the engineering opportunity on hold was the first of many difficult decisions Horwedel faced on his path to becoming a full-time fire fighter.

"In some ways it does make it easier to get a job if you have relatives in the fire service, but not the way people think. You still take the same tests and need the same qualifications as everybody else does, but you can speak the language and impress the chief."

The Los Angeles fire service is large, well organized, and competitive. Despite his numerous connections, Horwedel was in no way hired as a legacy.

"In some ways it does make it easier to get a job if you have relatives in the fire service," he explains, "but not the way people think. You don't really get any favoritism, but growing up around fire fighters makes it easier to give the chief a reason to hire you in your interview. You still take the same tests and need the same qualifications as everybody else does, but you can speak the language and impress the chief."

Horwedel had also had some hands-on fire-fighting experience. While he was in school, in addition to fill-in jobs such as working in a Chinese restaurant, managing a hotel, and working on a ranch, Horwedel spent summers on a "hand crew" in Los Padres National Forest near Santa Barbara. Fighting wildfires isn't classic fire suppression. Horwedel's job was to help cut a fire line around a wildfire by chopping and hauling away trees and brush in hopes of containing the blaze. It was tough, dirty physical labor, and kept Horwedel in good shape.

Testing and Hiring Process

So with family background and a first-hand familiarity working around fire on his side and a backup career path in place, Horwedel dove into the taxing process of applying for a job with the Los Angeles

City Fire Department (LAFD). The hiring process is long and complex, designed to screen out all but the most qualified and fit candidates for the academy.

The first step was clerical. Horwedel filled out and submitted a card stating his interest and was notified when the department administered its next written screening test. The test was something of a surprise. There were no questions about fire safety or fighting fire. The exam was all general knowledge—with math problems and even reading comprehension—more like an academic entrance exam than anything else. Applicants who pass the test are then granted an interview.

Horwedel took the exam with approximately 11,000 other applicants, 3000 of whom went on to have interviews. At the time, the LAFD didn't have many open positions, but the department kept everyone's scores on file. That one test was used as a hiring tool for quite some time. In the end, approximately 900 to 1000 candidates were hired out of the 3000 that passed the exam with Horwedel.

The next phase was the initial interview. Horwedel was quizzed on his background, interests, attitude, and ambitions. Interviews for a job with the LAFD are rigorous. The candidate is given an interview score; only those scoring 95 or above move on to the next hiring phase.

Next came a battery of medical, mental, and physical exams. Horwedel underwent psychological testing to determine his mental fitness to handle the stresses of fire fighting and, of course, physical exams to evaluate his physical fitness for the job.

The most demanding test at this phase in the hiring process involved performing a series of fire-fighting-related tasks to determine a candidate's endurance under simulated job conditions. Wearing firefighter jackets, but not full gear, candidates run what amounts to a timed obstacle course. The nonstop series of exercises includes dragging fire hose, dragging a human dummy, carrying a ladder, swinging a sledge hammer, pulling fire hose from ground level to a fourth floor, and crawling through an attic space. The course takes six to twelve minutes to complete; several recruits simply dropped out during the test.

"It was a demanding test," recalls Horwedel, "a real eye-opener. Fortunately, I was in good shape from my work with the Forest Service, and I passed."

The LAFD is almost obsessive about physical fitness. The department's newsletter for applicants (which features the motto: "You *can* stand the heat.") outlines specific training programs aimed at maximizing a fire fighter's upper body and other strengths. So, after surviving the field test, Horwedel was subjected to more physical testing that used physical training gear to measure his hand and leg strength.

Wearing fire-fighter jackets, candidates run what amounts to a timed obstacle course, including dragging fire hose, dragging a human dummy, carrying a ladder, swinging a sledge hammer, pulling fire hose from ground level to a fourth floor, and crawling through an attic space.

Then there were agility and endurance tests, for which Horwedel had to sidestep obstacles at full speed and prove how many times he could go up and down a flight of stairs in five minutes. These were pass/fail tests designed to eliminate all but the fittest recruits. Again, thanks to the physical workouts connected with his wildfire work, Horwedel passed.

Once Horwedel had proved himself through this battery of tests, he was assigned an investigator to carry out a background check. Investigators generally check applicants' arrest and conviction records and other character indicators. At the time Horwedel was applying for LAFD work, however, the primary background screening was to determine if applicants met the city's residency requirement, which has since been abandoned. Until recently, Los Angeles fire fighters were required to live within the city limits. Many candidates were eliminated for having given the city personnel department, which oversees the LAFD hiring process, a false address.

Throughout this grueling process, Horwedel had the full support of his family. "As far as my family was concerned," Horwedel says, "fire fighting was the only job I was going to do."

Once all the tests had been passed, Horwedel's records were sent to the city's fire chief for evaluation. The chief's office rates applicants according to their emergency medical technician (EMT) and fire-fighting experience in light of department needs. It then selects recruits for academy classes. Horwedel was still at California Polytechnic when he was selected to enter the LAFD's nineteen-week training program.

Training

"As far as my family was concerned, fire fighting was the only job I was going to do."

Horwedel trained at two locations, a facility in central Los Angeles and another near the city's harbor. Los Angeles is a huge city, packed with diverse ethnic and economic groups. According to Horwedel, his class at the academy—with its student mix of Hispanics, African Americans, Asians, and whites—represented a cross section of the city's population.

Los Angeles's academy training program is divided into four phases. The first, and least physically taxing, is devoted to EMT preparedness. This phase turned out to be the toughest for Horwedel; although he had gone to EMT school and had the certification, he'd never practiced dealing with medical emergencies.

Training phase two, laying hose, was strenuous, but Horwedel found it the easiest part of the academy. "I wouldn't say any of the training was easy, but learning and being tested on laying hose was the easiest part for me."

"I wouldn't say any of the training was easy, but learning and being tested on laying hose was the easiest part for me."

For the third training phase, Horwedel's class moved to the LAFD training facility near the Pacific Ocean for more physically demanding lessons. Here, the recruits learned the skills required to work on fire trucks. Students practiced handling and throwing (i.e., raising) ladders, over and over again. Horwedel, at 5 feet, 8 inches, (the fourth shortest member of his class) and 165 pounds considered this the toughest part of his training. "I felt I was amongst giants when I was at the academy," he says. "Those ladders made my shoulders sore and I was exhausted from carrying the weight, but I passed."

Having survived three phases of training, Horwedel's class moved on to live-fire practice, during which they fought real fires set in a six-story cement training building called the tower. For several weeks, trainers would set blazes in various parts of the tower and students would respond, using the skills they'd mastered earlier on—laying hose, throwing ladders, and attacking the fire.

This final training course also introduced the cadets to other, more specialized, aspects of fire fighting. Members of the hazardous materials (*hazmat*) squad, a specialty division within the LAFD, gave presentations to teach cadets what's expected of fire fighters when they interact and cooperate with hazmat experts at a fire scene. The class also encountered urban search-and-rescue techniques, such as rope training, and participated in simulated river rescues.

Perhaps one of the most dramatic, and ultimately most vital, lessons came out of a field trip back into the city. Horwedel and his classmates were taken to downtown Los Angeles where they threw a ladder and climbed to the roof of a building. From there, instructors guided them on a tour of urban rooftops, pointing out the various kinds of roofing materials and the structural aspects of roofing on commercial and residential buildings. As Horwedel was to learn firsthand during his rookie year, roof access and ventilation are a major part of fighting a fire. Studying architecture and construction techniques prepares fire fighters to work on top of a burning building safely and with confidence.

Roof access and ventilation are a major part of fighting a fire. Studying architecture and construction techniques prepares fire fighters to work on top of a burning building safely and with confidence.

Toward the end of academy training, recruits spent a night in a real fire station where their sleep was interrupted by a series of simulated emergency calls—a car fire, a rubbish fire, and a (smoke-only) structure fire involving simulated trapped victims.

Even for the physically-fit Horwedel, the academy experience was demanding to say the least. Training required his full physical and mental concentration, and Horwedel got his first real taste of the price he'd be paying through his rookie year when he realized how much he was neglecting his family and his other interests for the sake of getting the most out of his studies.

Throughout the 19 weeks at the academy, cadets participated in strenuous daily physical training: running, calisthenics, chopping logs, and pulling hose. Even Los Angeles's experienced fire fighters are not required to perform fitness routines as often or as rigorously, although the department has implemented a wellness program under which fire fighters are encouraged to keep in shape with daily exercise.

Because cadets are so thoroughly screened before they're even put on the list of applicants to be called to the LAFD academy, the survival rate for recruits is high. During Horwedel's training, no one quit the academy. A few were mustered out because of their inability to meet standards during hose or ladder training. Those high standards made the experience all the more successful for Horwedel and the rest who made the grade.

The academy atmosphere was militaristic. Trainers were unforgiving. A recruit would be taught to throw a ladder one day and tested on his ability the next. "It was a lot of pressure," says Horwedel. "When you were tested, you knew you were throwing that ladder for your job." Most recruits did fail the ladder throw on their first test but passed on their second (and only) retest. Ultimately, the academy's

"It was a lot of pressure. When you were tested, you knew you were throwing that ladder for your job."

pass-or-fail ethic instilled confidence in the majority of cadets who stayed the course. Horwedel left the academy fit and trained for his probationary year.

Rookie Year Experience

Horwedel started his rookie year on September 28, 2001. In the LAFD, the months a recruit spends at the academy don't count toward his or her probation. It was at this point that Horwedel's career training truly isolated him from the rest of his life.

"Probation is a sacrifice," Horwedel explains. "It was difficult for me and for my family because I became totally involved. I had no time for anybody else or any other activities, but it was a necessary sacrifice. When you're training for the fire service, you have a responsibility to learn the job because it's dangerous. You're obligated to learn to do it right so you can come home to your family after every shift."

Horwedel's probationary year was divided into three station assignments planned to expose rookies to working on a truck, riding with an engine company, and running EMT responses, in that order. The LAFD designs these training stints so that rookies are always partnered with an experienced fire fighter.

Horwedel's first rookie assignment was to a "slow" station, Station 85, in Harbor City, not far from the academy facility where Horwedel had trained. Station 85 maintains a "task force": one truck, a pumper, and an engine. He was assigned to work the "top man" position, a job that kept him under the constant supervision of the apparatus operator (driver). As top member of the truck squad, Horwedel's job was to throw a ladder up to the roof of a burning structure as soon as the truck arrived at a fire scene. The trick is to choose the proper ladder and then, with the help of the tiller man (the fire fighter who steers the rear of a ladder truck), throw it against the building as quickly as possible. The top man and the apparatus operator then climb to the roof. The apparatus operator cuts a hole in the roof to allow smoke to escape the building. The top man removes roofing material as the apparatus operator cuts.

"Probation is . . . a necessary sacrifice. . . . You're obligated to learn to do it right so you can come home to your family after every shift."

"During your first rookie station, you really learn roofs," Horwedel explains. "You climb them all—flat, pitched, arched—and they teach you to be safe up there."

The top man's job is to throw a ladder up to the roof of a burning structure as soon as the truck arrives at a fire scene, climb up, and remove roofing material as the apparatus operator chops a hole in the roof to let the smoke escape.

Horwedel's first call at Station 85 was something of a letdown. His was the second truck to arrive, and by the time he got there, the fire was mostly under control. Going by the book, he threw a ladder and climbed to the roof, but a crew from a truck equipped with an aerial ladder had already done the rest of the job. Still, any experience with a live fire is valuable, and Horwedel came away from each call a more seasoned rookie.

"I didn't learn everything." Horwedel admits. "Even guys who have been with the department for years say they learn every day. But during my rookie year, they gave me the chance to learn a lot of skills."

At Horwedel's second station assignment, at LAFD Station 91, Horwedel's captain pushed him to pick up as much hands-on experience as possible, including learning to drive the apparatus and pump. His primary assignment, though, was to act as the nozzle man. The fire fighter in that position is always teamed with the engine's captain.

Station 91 was another small firehouse. It ran only one engine and a paramedic ambulance, but it was busier than Station 85 had been. In addition to responding to many car accidents and car fires that were often set in abandoned stolen vehicles, Station 91 was also called to a greater number of structure fires.

The nozzle man has a high-profile job. Once the hose is hooked up at a fire scene, the nozzle man sets up a line to the fire, puts on breathing apparatus, and enters the burning building. It was during this rookie rotation that Horwedel learned to appreciate the value of the top-man job he'd practiced at his first assignment.

"I knew what to expect. We were taught why you cut the roof and what would happen, but it was amazing to see the theory in practice from the inside," Horwedel says.

"During your first rookie station, you really learn roofs. You climb them all—flat, pitched, arched—and they teach you to be safe up there."

Once the hose is hooked up at a fire scene, the nozzle man sets up a line to the fire, puts on breathing apparatus, and enters the burning building.

Horwedel's third rookie assignment was to Station 26, a mid-city station situated between downtown and southcentral Los Angeles. The area is characterized by wooden Victorian homes dating from the late 1800s, commercial strip malls, and many standard "center hall" apartments (notorious as the scenes of high-casualty fires). There are only two high-rise structures in the neighborhood, much of which houses the poor and disadvantaged. There are a high number of violent crimes—shootings and stabbings—in the area, but, Horwedel reports, the residents have a generally good attitude toward fire fighters. The area covered by Station 26 is ethnically diverse, with a strong Korean community to the north, a large Muslim population, and many Hispanics.

At Station 26, Horwedel was meant to train on an ambulance with the EMTs. But Station 26 was a very busy station, and Horwedel soon found himself filling top-man and nozzle-man slots. Eventually, he rotated through the other engine positions—tillerman, hydrant man, and inside man. It was a tough assignment, but it was excellent training.

Such a high degree of fire training is unusual. An average firehouse in Los Angeles will get ten to fifteen medical emergency calls a day; whereas, thanks to smoke alarms, sprinkler systems, and modern prevention methods, calls to live fires are relatively rare. Still, Horwedel notes, the fire calls are the dangerous part of the job, and the fact that you get less on-the-job experience fighting fires makes fire training all the more valuable.

A year after Horwedel left the academy, he became a full-time fire fighter, assigned to Station 26, the last station where he'd trained. He is qualified to work all positions available at that station—truck, engine, and Basic Life Support ambulance.

Horwedel finds that being on the job is very different from working on probation. In the Los Angeles Fire Department, rookies have real assignments. They're an essential part of the firehouse teams, not observers. With the end of the probationary year, however, comes more independence. New full-time fire fighters do the same jobs but they're not always paired up with someone more experienced.

"When you go off probation," says Horwedel, "you join the fraternity. You've proven yourself. As a rookie, you were tested every day. Now you're trusted and accepted, and you do more things."

But the hard work never ends. The day Horwedel got off probation, he was only a little bit relieved. "You can't relax," he explains. "Even through there's no training schedule, I know that I'm young and I don't know much."

"When you go off probation, you join the fraternity. As a rookie, you were tested every day. Now you're trusted and accepted, and you do more things."

During his rookie year, Horwedel admits, he was the one pushing him the hardest. It took Horwedel at least two months after probation to relax a little, and, as the pressure lifted, he started having more fun—on and off the job.

7

Stress Tested

Jesus Moreno

Jesus Moreno

PROFILE

Fire Department: Dallas Fire Department

Location: Dallas, Texas

Position / Title: Fire fighter

Rookie Year: 2002

Previous Job: orderly, Baylor Hospital, Dallas, Texas

Education / Training: Paramedic training, Methodist Central Hospital

Reason for Joining the Fire Service: "Couldn't stand working indoors and was attracted to the excitement and challenge of fire fighting."

Most Memorable Rookie Experience: Working inside a burning building, holding a hose and then suddenly realizing that his feet and legs felt hot and looking down to see flames blowing through his legs.

Fire Service Ambition: Continuing to do what he's doing. "At times, the job is rough, but I couldn't think of doing anything else. They shouldn't even pay us to do this."

Jesus Moreno grew up a few blocks from Station 26, where he now works as a full-time fire fighter with the City of Dallas Fire Department. When Moreno was in grade school, his class took a field trip to Station 26. Back then, it was an older building, replaced by a modern firehouse in 1994, and it had a pole for fire fighters to slide down from their quarters to the truck bays. Naturally, Jesus's class was warned to use the stairs, but some of the boys, including Jesus, slid down anyway.

Family and Background

For some, this childhood experience might have been the beginning of a lifelong dream of joining the fire service, but for Jesus, it was just a prank. He'd never thought of being a fire fighter before, and it didn't cross his mind again until decades later when, while in his mid-twenties, Moreno was unhappily working at Dallas's Baylor Hospital, doing orderly, secretarial, and basic patient-care tasks.

"I couldn't stand working indoors," Moreno explains. "I've always been an outdoor person, and I didn't care for working inside the hospital."

Moreno's hospital duties included stints in the ER, where he met Dallas fire fighters as they brought in crisis cases. It was through these EMTs that Moreno learned how to apply to the fire department.

"I knew I wanted to do something else," Moreno recalls. "I was impressed by the Dallas fire fighters, but I didn't really understand the details of their jobs. I thought fire fighters just rode on the trucks and put out fires. I had no idea about operating procedures—no idea how complex the job was."

"I thought fire fighters just rode on the trucks and put out fires. I had no idea about operating procedures—no idea how complex the job was."

Since his initial exposure to the Dallas department was through EMTs, Moreno began his career move by enrolling in paramedic courses at Methodist Central Hospital in Dallas. He managed that year of paramedic training on top of a 60-hour workweek. "I slept whenever I could, and that wasn't very much," he says. Somehow, however, he did find time for a social life; and during that year, he met his fiancée, Kristen, who was in nursing school at the time.

Moreno has no family background in fire fighting. His father did radio repair and his mother worked for a printing company. Aside from a brother in the Navy, who took fire-fighting instruction as part of his aircraft-carrier training, nobody in Moreno's life had had any connection with a fire department. Still, it was his family upbringing that prepared him for fire fighting—that and an attraction to excitement.

"I was brought up in a very strict home," Moreno explains. "I was used to rules and used to the idea that everything had to be done a certain way. Some people find the discipline of the fire department a little hard to take, but having all the rules seemed normal to me."

"Some people find the discipline of the fire department a little hard to take, but having all the rules seemed normal to me."

Moreno also favors another aspect of fire fighting that many recruits find difficult. "I decided that if I couldn't work for the Dallas Fire Department, I didn't want to work for anybody else," he says. "I know this sounds backwards, but it's an edgy department. There's a lot of tension. People get agitated. Sometimes it's abrasive because we have a busy city with busy stations. It's run, run, run all day—very high stress. It's sink or swim. You don't see that challenge at other departments."

Testing and Hiring Process

Moreno's path to becoming a fire fighter began with an extensive application process designed to weed out all but the most dedicated would-be fire fighters. The first step was a City of Dallas civil service exam, which included 1000 questions: 500 to measure aptitude; 500 to measure psychological fitness. Applicants who pass that test face what Moreno describes as the toughest thing he'd ever encountered, the department's physical agility test.

"The agility test weeds out a lot of people," explains Moreno. "I thought I was in good shape. I could run, and I could lift heavy weights, but that test kicked my ass."

The test, with an emphasis on lower-body strength and endurance, is administered twice on Fridays—once in the morning and once late in the day. Because the test is so physically taxing, it's canceled if the temperature goes above 92 degrees or if it's raining.

Before deciding to focus his ambition exclusively on the Dallas Fire Department, Moreno had looked around at other local fire services—specifically one in a nearby community called Grand Prairie—so he'd taken and witnessed other fire fighter fitness tests. Compared to those, the Dallas test looked easy. To look at it, Moreno never thought the Dallas agility routine would be as grueling as it was. He was unpleasantly surprised.

"Three-quarters of the people who finish the test can't walk for five or ten minutes," says Moreno. "During the test, somebody usually passes out. Often an ambulance shows up. The exercises test hand-eye coordination, overall agility, and especially lower-body strength. But it escalates—as you get tireder, the jobs get harder."

"The exercises test hand-eye coordination, overall agility, and especially lower-body strength. But it escalates—as you get tireder, the jobs get harder."

The test is a series of timed tasks, including running on a roof, re-racking a ladder, dragging a 1¾-inch hose 50 to 75 feet, running up 4 stories of stairs while holding a 45-pound bag at waist level, and, ultimately, moving a sopping wet, charged 3-inch hose, in three 50-foot sections, 25 feet in each direction.

Examiners establish a baseline time for each test by having more experienced cadets run the course. So the maximum time fire fighter applicants try to beat will vary from day to day to compensate for outside factors such as humidity and temperature.

The Friday that Moreno took the test, the target time was 323 seconds (roughly 5.5 minutes). The test is administered in segments, with a minute (untimed) "rest" period between tasks. Moreno completed the test in 291 seconds (just under 5 minutes).

"That was good, but I'd actually expected to do better," Moreno recalls. "I was 180 pounds at the time, and I ran 5 miles every other day and lifted weights. But when I hit the last part of that hose drag, it was next to impossible to move."

Out of the ten applicants who were tested with Moreno, six passed. The other four dropped out. The test is so taxing, that it's common for people to quit—sometimes just feet away from finishing—even though applicants have to complete the test to be eligible to take it again.

"I don't know why some people get that last hose halfway and still give up," Moreno wonders. "If they finish the test—even with a bad time—they could try again, and still they drop out. But I think it's the right system because once you get to the academy you see that the instructors care less about how fast you do things as they care about the fact that you don't quit. They really want to make sure you're psychologically ready for the job. They don't want fire fighters giving up in a dangerous situation."

"Instructors care less about how fast you do things as they care about the fact that you don't quit. They don't want fire fighters giving up in a dangerous situation."

Once he'd survived the physical test, Moreno was subjected to an extensive background check. The city checked all his previous employers and recent addresses and had him complete a personal history packet that asked a series of job-related and personal questions. The information Moreno gave was then used for a polygraph test during which he was questioned on everything from the largest amount of money he'd ever stolen (there's a cutoff number above which a candidate is considered unacceptable) to his sexual history.

Moreno's only discrepancy on the polygraph involved, ironically, his name. "Is your name Jesus Moreno?" the test-administrator asked.

"Yes," Moreno answered, but the polygraph registered that he was lying. "I think I was nervous," he says now. "That was an easy one to straighten out."

At that point, the final obstacle between Moreno and the academy was an interview with three Dallas deputy fire chiefs. "They make you sweat and they put you on the spot no matter what your record is. I'd talked to a lot of other candidates about their background checks, and, compared to some of them, I looked like an angel. But the chiefs are determined to find something to pressure you with. They want to know how you respond under stress. I think the department figures that if they're going to hire you for some 20-odd years, they want to be sure they're going to get those 20 years out of you."

Moreno passed muster and was given an offer of employment, his ticket to the Dallas fire academy. His friends and family supported his decision to go into fire fighting, even though it meant putting his personal life on hold for the next nine months.

Training

The Dallas fire academy offers the traditional mix of classroom study and physical training, but it was the fitness instruction that made the biggest impression on Moreno. It turned out that the agility-screening test was just a warm-up. The academy's approach was paramilitary, which Moreno, harking back to his childhood upbringing, found comfortable. "I've been known to be difficult, but I have to respect someone who isn't taking any of that," he confesses.

Moreno had looked into fire training at local community colleges but determined that, although the community colleges taught basically the same procedures and techniques, they didn't prepare cadets for the physical extremes of the job. During his academy training, Moreno injured his knee and pulled a muscle in his thigh, but decided it was better to tough it out than report his pain.

"A lot of guys did that," he says, "because if you report an injury and have to go out, you come back in at the beginning of the physical training process. So people just ignore their injuries. I don't know anyone who would want to go through that again. The physical training at the academy beats you down. They're always pushing you. The department's philosophy seems to be that if you're going to fail, be carried out. Don't quit."

The rigors translated into a strong bond among the cadets, and Moreno admits that it's hard for him to respect a fire fighter who didn't train at the Dallas academy. "You can learn from books to pass the state test, but that's just theory. You learn more in the field. It's experience that establishes the pecking order within the department. At the academy you aren't even a probationary fire officer, so you learn to respect people with seniority—people who've done things the way the Dallas department really does them every day."

The fire-fighting course at the academy lasted nine months. Because he'd had extensive paramedic training, Moreno was able to skip some of the EMT curriculum and take a bridge course that trained him on Dallas ambulances so he could learn the department's specific procedures.

Moreno did ten EMS shifts and was graduated from the academy the day after his tenth shift. Two days later, he began his rookie year, temporarily assigned to Dallas Station 32, where he'd done his cadet ambulance runs.

Rookie Year Experience

The break-in system was equally swift. For their first ten shifts, Dallas rookies ride on an engine with backup and have their abilities and experience evaluated by veteran fire fighters. After that, rookies work regular shifts, half with the ambulance and half with the fire engine, and are given the opportunity to do just about everything. It's a true "baptism of fire."

"You can learn from books to pass the state test, but that's just theory. You learn more in the field."

Moreno's first fire call, at Station 32, was anticlimactic but instructive. As the truck Moreno was riding pulled up, the fire fighters could see that the house was fully involved. In fact, all but one room was ablaze, as was the garage and a tree behind the house. The fire had vented itself by burning away the building's roof. Thankfully, there was no one inside the house, thereby eliminating the need for conducting a search or rescue.

However, Moreno, following procedure, pulled a hose line up to the front door and learned a valuable lesson—namely, that with the combination of their masks and the presence of smoke, fire fighters often work in the blind. In this case, nobody had to enter the burning building. The crew's job, instead, was to contain the fire and extinguish it from the outside.

On another call to the same neighborhood, Moreno did enter a burning structure and came away with another lesson that the academy couldn't teach. "You don't realize how well your gear works—sometimes until it's almost too late," Moreno explains.

Moreno was working inside the burning building, holding a hose and feeling fairly comfortable, when suddenly he realized that his feet and legs felt hot. He looked down to see flame blowing through his legs. "You have to learn what your gear will tolerate. It's a lot to get used to," he says.

Moreno got a second real-life lesson on that subject at a call to a smaller house. He went inside with a hose. He stepped in further and his foot went through the floor. Then his other foot went through, and he realized that the entire wooden floor was glowing red—like the embers in a fireplace. "The gear works so well," Moreno says, "that you have to pay attention. It's too easy to go in too far."

After more than three months at Station 32, Moreno was reassigned to Station 26, in his old neighborhood. For Moreno, it was a dream-come-true. From the start, he'd wanted to work at the station near

"The gear works so well that you have to pay attention. It's too easy to go in too far."

where he'd grown up, but Dallas has 54 firehouses and Moreno assumed landing at the one he preferred was a long shot and might never happen. When he was reassigned to Station 26, he couldn't believe his luck.

Dallas Fire Department Station 26 is a very busy double-company station. It maintains a rescue truck, a fire engine, and an ambulance and works ten fire fighters to a shift. The station serves Oakcliff, a rundown area of Dallas, next to an old-money section of the city. The neighborhood has a reputation for violence, and the station responds to an above-average number of fire calls.

As is the case at most fire stations, the ambulance side is the busiest at Station 26. When the weather's cold, according to Moreno, you can expect up to 18 calls during a 24-hour shift. When it's hot, the number can jump to 24 or even 30 calls. At Station 26, ladder trucks respond to freeway accidents, primarily to block traffic and protect the EMTs.

Again, following nationwide trends, house-fire calls in Dallas are on the decline, with a heavy period involving one per week. Car fires are much more common. Some of those are stolen cars that are torched; others accompany traffic accidents in which air-bag cartridges overheat upon discharge; still others are the result of malicious acts. "We had one incident when a woman got mad and torched her boyfriend's car and his mother's car," Moreno recalls.

The busy atmosphere at the station suited Moreno, and gave him ample opportunity to pick up experience. As a rookie, he also learned how much work goes in to running a fire station. "When you're a rookie," he explains, "they expect you to be the one who jumps up and answers the phone. If something needs to be cleaned, they expect you to be the first one on it. And you can pretty much count on being the one who cleans the restrooms." But at the same time, Dallas rookies are being evaluated constantly for their professional performance. Veterans offer feedback and officers follow up calls with corrective direction. As at the Dallas academy, there's a heavy emphasis on rules and procedures, which Moreno finds comfortable.

"We ran mock drills at the academy, but nothing is as physically demanding as live fire fighting."

Moreno's rookie months were also a time for little lessons—like learning not to eat too much at one time. "I always had a fast metabolism," Moreno says, "and I never worried about gaining weight or how much I ate. But one time I ate a lot at the station just before I went out in the field, and I threw it all up. Later I talked with some of the veteran fire fighters and they told me that was actually a common experience. Now I make sure I eat smaller meals."

Six months after his probationary stint has ended, Jesus Moreno looks back at his training time with a lot of pride and a certain degree of relief. He's found fire fighting to be more physically demanding than he'd imagined and learned that, thanks to heavy gear and high heat from fires, a fire fighter wages a nonstop war with dehydration. "We ran mock drills at the academy," Moreno explains, "but nothing is as physically demanding as live fire fighting. Since I left the academy, though, my knee and thigh have healed up, so I perform better now than I did at the beginning."

Because Station 26 serves a heavily Hispanic area of Dallas, Moreno's own Spanish is improving rapidly. It helps, he says, to understand the culture of the community and to recognize the danger signs as well as to know how to get along.

Being Hispanic in a major city with a huge Latino population and a fire department with currently less than 10 percent Hispanic fire fighters was less of an issue than Moreno had expected. "Being a minority is part of my life," he explains. "Over a lifetime, you learn to work with it and around it. It's no different here at the fire department. Some people bring their problems with them to work, but you know that you're going to be depending on them in the field, so you let things go."

"Being a minority is part of my life. Over a lifetime, you learn to work with it and around it. It's no different here at the fire department."

Race, Moreno contends, becomes secondary to the fact that fire fighters share experiences and perspectives that civilians can't imagine "What I have learned is this," Moreno continues. "When you're a fire fighter, you see people and things that go on in the city that most people never realize exist. It makes you kind of a recluse. I tried describing some of the things I saw and some of the situations I was in to my family, and after a while, they told me to stop. There are things you can't discuss with the general public. But other firefighters always understand. Other firefighters 'get it.' "

The daily stress and exertion haven't discouraged Moreno in the least. "There are no bad firehouses," he says, "just slow ones. At times,

"If you fit the job, it's your home."

the job is rough, but I couldn't think of doing anything else. They shouldn't even pay us to do this. If you fit the job, it's your home."

A Second Career

8

Arthur Moy

Arthur Moy

PROFILE

Fire Department: Engine 9, Cambridge Fire Department

Location: Cambridge, Massachusetts

Position / Title: Fire fighter

Rookie Year: 2001–2002

Height / Weight: 5′7″ / 170 lbs.

Previous Job: Civilian contract engineer with United States Department of Defense/USAF

Hobbies / Other Interests: Flying airplanes and helicopters, scuba diving (wreck/tek and cave), rock climbing, skydiving

Education / Training: B.S. electrical engineering, Northeastern University

Reason for Joining the Fire Service: Opportunity to do important and significant work

Most Memorable Rookie Experience: First fire

Fire Service Ambition: "My daily ambition is to gain experience; learn new skills; and avoid complacency, poor judgment, and mistakes. My long-term ambition is promotion."

Arthur Moy showed up even later than expected for his fire fighter training, but he had a good excuse. Moy was appointed on Sunday, September 9, 2001, and told to report to training for the Cambridge, Massachusetts Fire Department on Monday, September 10. That weekend, however, Moy was halfway around the world on Bikini Atoll in the Marshall Islands. He was on vacation, diving to the wrecks of World War II battleships that were sunk off Bikini during America's 24 post-war atomic and hydrogen bomb tests conducted between 1946 and 1958.

Moy, an avid scuba diver, had been planning the adventure trip for a year and a half and had gotten special permission from the Cambridge Fire Department to arrive for preliminary training sessions on September 12. Full-time fire-fighting academy instruction wasn't scheduled to begin until September 17.

It was on this remote group of Central Pacific volcanic islands, 1200 miles east of Guam, that Moy first heard the news of the planes crashing into the Pentagon and World Trade Center towers on September 11, 2001. Word came over the AM radio in a pickup truck.

Moy didn't learn the details or the extent of the terrorist attacks until he reached Majuro Atoll, capital of the Marshall Islands, to begin his flight back to the states. By then, air traffic to the United States had been suspended, and Moy found himself stranded very far from Massachusetts. As a result, Moy didn't arrive at the academy until September 20. He had missed his initial training and orientation week in Cambridge and three days of his time at the academy (just surpassing the academy's absence limit). Because of the extraordinary times and circumstances, the department allowed him to continue his enrollment on the condition that he have no more absences.

Background

Fire fighting is not Arthur Moy's first career. At age 45, he is one of the older "young fire fighters" in the Cambridge department. Moy's first career was electrical engineering, which he studied at Northeastern University in Boston. As part of Northeastern's co-op learning program, Moy worked for the Department of Defense at Hanscom Air Force Base in Bedford, Massachusetts, northwest of Boston. After graduation, Moy stayed on as a civilian contract engineer, working primarily with radar and electronic materials.

By the end of the century, the Air Force was downsizing, bases were being realigned, and the lab where Moy worked was periodically threatened with a shutdown. Layoffs were taking place, and workloads for the surviving civilian employees increased. Everything was uncertain. With the change in work atmosphere, Moy began considering a move.

A scuba-diving friend of Moy's had become a fire fighter with the Cambridge Fire Department and suggested that Moy take the screening exam. Moy first took the test in 1997, but there was no opening. At his scuba-diving friend's encouragement, Moy took the test again in 2000. This time the department was hiring.

Still unsure if he was serious about a career in fire fighting, Moy filled out an application and underwent initial interviews. His interest grew, especially because the once-booming New England high-tech economy was turning sour. Engineers who left full-time employment were no longer able to write their own tickets as consultants.

Further, Moy's sister had recently abandoned a career as a lawyer to join the FBI. She lit a fire under him, says Moy. "She told me she was having a much better time with the FBI and encouraged me to take the plunge. She warned me that it would be hard to get an offer, but that if I got one, I should take it."

"[Fire fighting is] a good profession for a gearhead. I like to play with toys and equipment."

Before that, Moy's interest in fire fighting had been casual. Like most children, he saw fire fighters on TV and loved the look of the adventure. "It's also a good profession for a gearhead," Moy explains. "I like to play with toys and equipment."

Moy is fit and fond of adventure. He flies small airplanes and helicopters, cave dives, rock climbs, and skydives. Taking all these factors into consideration, fire fighting seemed like a realistic career change. After Moy passed the department's physical ability test, the Cambridge Fire Department offered him a job. He accepted.

Moy says that despite his age, he had no trouble with the physical ability test, a series of timed events that included climbing stairs while wearing a weighted vest, dragging a hose and a weighted sack through a crawl maze, shuttle running with a small ladder, navigating a blackout maze, pulling the pike-pole on a ceiling overhaul simulator, and moving a tackling sled with strikes from a sledge hammer. It was a pass/fail trial, and Moy passed with relative ease.

Training

After almost having his career change sidetracked by the tragedy of September 11, Moy reported to the Massachusetts Firefighting Academy in Stowe, Massachusetts. Run by the state government, the Massachusetts Firefighting Academy is a centralized training facility to which any fire department in the Commonwealth of Massachusetts can send its recruits for training, at no cost to the local community. The Cambridge Fire Department requires that all of its fire fighters attend the academy, a policy that enables the department to maintain its ISO (Insurance Service Organization) Class One status.

The standard eleven-week training program at the Massachusetts Firefighting Academy consists of both classroom and practical instruction. The academy, which prides itself on its students learning "more than how to put the wet stuff on the red stuff," also offers hundreds of classes for professional development each year.

Cambridge fire fighters are 100 percent trained to use a defibrillator. Although it's not required, a high percentage of Cambridge fire fighters are EMT certified, with many working as EMTs or paramedics in side jobs. Because the academy focuses on fire service training, the Cambridge department will reimburse fire fighters for their separate EMT training.

Moy's probationary year officially began with his admission to the academy. Although his training took place in the shadow of the terrorist attacks of September 11, Moy contends that the recruits were working too hard to be distracted. Still, Moy's class did not have a totally typical academy experience. For one thing, at each morning's roll call they observed a moment of silence in memory of the 343 New York fire fighters missing at the World Trade Center.

Also, despite the trainers' efforts to keep things on a business-as-usual basis, fire fighting had suddenly become an extremely high-profile occupation. Even though the academy instructors stuck to the standard curriculum, it was impossible not to be haunted by current events when the classes covered high-rise evacuation procedures, building collapses, body searches, and rescue-and-recovery efforts. Some of the academy instructors, Moy remembers, went down to New York to help with the dig-out and returned with first-hand accounts.

Solemn overtones aside, Moy claims he had more fun at the academy than he has on the job with Cambridge Engine 9 because the last five weeks of training involved fighting 10 to 15 simulated fires. With all his adventure sports training, Moy did not find the physical aspects of the academy difficult. He notes that the class always worked as a group—running in formation, doing exercise drills together—so that nobody was ever left behind.

Much of the practical fire-fighting training at the academy involved repeated evolutions of simulated standard procedures during which recruits were scored on a demerit system. Instructors would evaluate aspects of a student's performance and award 1 to 5 demerits for each minor error or each safety error. If a recruit showed up for an exercise without his face-piece, a safety issue, the recruit would earn the maximum of 5 demerits. A less serious mistake—say, a twisted strap—would cost the recruit just 1 demerit. Demerits can accumulate quickly, but the focus of the demerit system was to make students aware of *not* making mistakes. Moy recalls that the demerit limit was about 124, but the worst score in his class was only around 60, with the average in the 20s. The "golden boy" of the class (not Moy) left with only 2 demerits.

Moy's favorite part of the academy experience was the simulated structure fires. Recruits would carry bales of hay into one of the academy's concrete live-fire buildings, then they would leave so they wouldn't see where the instructors arranged the fuel. The live-fire simulations were realistic and thorough runs—all the way from receiving the call at a station to cleaning up the equipment after the fire was put out.

The live-fire simulations were realistic and thorough runs—all the way from receiving the call at a station to cleaning up the equipment after the fire was put out.

"One of the things they taught us was to always size up each incident by paying attention to radio chatter (from the fire alarm and while en route), so we could step off the apparatus with situational awareness of the scene," explains Moy. "because when you get to the fire, heavy equipment, smoke, urgency, and noise interrupt fire-fighter communications and create confusion. The challenge is not to let confusion become mass chaos."

Rookie Year Experience

After graduating from the Massachusetts Firefighting Academy on November 30, 2001, Moy went to work with the Cambridge Fire Department. The Cambridge department maintains a specialized training

"One of the things [the trainers] taught us was to always size up each incident by paying attention to radio chatter, so we could step off the apparatus with situational awareness of the scene, because when you get to the fire, heavy equipment, smoke, urgency, and noise interrupt fire fighter communications and create confusion."

division that provides rookie indoctrination and performs ongoing training and evaluations (including quarterly evaluations of rookie fire fighters) throughout the department. Moy was assigned to Cambridge FD Training Division, where, for the next several weeks, he was trained in rules and regulations, standard operating guidelines (SOGs), first-responder and CPR certification, defibrillator operation, the incident command system, for example. Moy was also given driver training on the department's fire engines—something the academy does not offer.

With a resident population of 101,000, Cambridge, Massachusetts, which occupies 6¼ square miles directly across the Charles River from Boston, is the sixth densest populated city in the United States. Its daytime population swells to more than 400,000. The city is home to several colleges and universities, including Harvard University and the Massachusetts Institute of Technology (MIT). Business and shopping districts surrounding the schools line congested thoroughfares that are flanked by narrow residential streets. Traffic and parking are serious problems, and much of the city's housing includes old wood-frame multifamily buildings. Because of the real danger of a fire spreading among the city's closely packed homes, the Cambridge Fire Department's primary mission, according to official policy statements, is to contain structure fires to the building of their origin.

Fortunately, thanks to a combination of factors, including a series of real-estate booms over the last decades of the twentieth century that resulted in housing units changing hands often, the number of structure fires in Cambridge is in steady decline. Every time a Cambridge property changes hands, it undergoes inspection and is brought up to fire code specifications. In the years 2000–2001, the Cambridge Fire Department made 30,219 runs citywide, but only 319 of those were classified as building fires.

The 274-member Cambridge Fire Department is well equipped, with a full complement of engine and ladder trucks plus heavy-rescue vehicles, all-terrain vehicles, hazmat (hazardous materials) response vehicles, and marine equipment to patrol the riverfront and the

Charles River Basin.

Cambridge also contains some of the most expensive single- and multifamily houses in Eastern Massachusetts. Many of these homes are located in West Cambridge, a primarily residential area protected, in part, by Engine 9, to which Moy was assigned for the remainder of his probationary time with the department

"There are friendly rivalries among the department's fire companies. We get ribbed at Engine 9," says Moy. "They call us a retirement home because West Cambridge doesn't get as many calls as the business districts or the more densely populated areas of Cambridge. Our running card, however, sends us to 80 percent of the city's second alarms," explains Moy. Engine 9, therefore, ends up going to most of Cambridge's sizeable fires, inspiring Engine 9 drivers and officers to respond to the friendly ribbing with the rejoinder, "At Engine 9, we do *quality,* not *quantity* calls."

Because of the real danger of a fire spreading among the city's closely packed homes, the Cambridge Fire Department's primary mission, according to official policy statements, is to contain structure fires to the building of their origin.

The mix of emergency calls to the fire service has changed. Although the numbers of fires across the country has declined, EMS calls and false alarms have increased. Most recently, the fire service is branching out into terrorism emergency response. Cambridge Engine 9 cannot escape this trend. In 2000–2001, the station logged 981 responses, of which 29 were building fires, 362 were medical emergencies, and 18 were mutual-aid calls to the bordering towns of Boston, Brookline, Belmont, Somerville, Waltham, and Watertown. In addition, Engine 9 responds to all second alarms throughout the city of Cambridge.

At Engine 9, Moy's captain assigned him to work with one of four groups of fire fighters providing 24 / 7 fire duty for West Cambridge The fire fighters within each group are cross trained and interchangeable at the pump, nozzle, and hydrant positions, though Moy allows that individuals develop preferences and (at the captain's discretion) often a group works together best by having unofficial "fixed" duties.

Although he was just out of the academy, Moy's rookie status didn't relegate him to being an extra hand. Instead, he was dropped right into action, expected to perform jobs right along with the company's experienced fire fighters.

"While you're on probation, people do treat you differently," says Moy. "They make a little fun of you. You always know that you're the new guy. You undergo quarterly performance evaluations and you know that you're on probation—that the department can let you go at the end of the year if you don't work out."

"While you're on probation, people do treat you differently. [. . .] You always know that you're the new guy. You undergo quarterly performance evaluations and you know that the department can let you go at the end of the year if you don't work out."

It was three months before Moy went to his first fire. The incident began as a mutual-aid call to nearby Watertown. The midnight fire in a duplex house went from one to three alarms, with the entire Watertown department on the scene. Moy and the crew from Cambridge Engine 9 went to a Watertown firehouse to cover that station in case another call came in. As the fire escalated, the Watertown crews called for more help on the scene. By this time, the incident commander on the scene had moved fire fighters out of the building and declared the fire an exterior operation. Earlier in the evening two young women had escaped the building with burns. An elderly man had died inside.

Engine 9 approached the duplex from the rear and ended up on the front lawn, pumping supplemental water to the engines already operating and handling attack lines to hold the fire at bay from one side (to protect exposures). Before it was over, there were trucks from Cambridge, Boston, and several other towns on the scene.

Engine 9 also protects a large park area surrounding West Cambridge's Fresh Pond Reservoir, so the company deals with something urban firehouses seldom see—namely, brush fires. As is the case with most modern fire stations, however, the majority of the calls Cambridge Engine 9 responds to are medical. Moy has seen his share of car accidents. Even though Cambridge has extensive medical and rescue services and contracts with private professional medical teams to cover the citywide volume, it is often the life-sustaining first-aid administered by neighborhood stations that saves lives.

Because Cambridge is geographically small, the large medical units are always relatively close, but heavy traffic can delay response. The Engine 9 EMTs, armed with oxygen tanks, a defibrillator, medical equipment, and CPR expertise, can arrive at a local emergency scene first, size up and stabilize the incident, and provide first response to keep a victim alive until basic and / or advanced life support transport apparatus arrives.

This is exactly what Moy's group did in the case of a drug overdose victim who had collapsed on the street and was turning purple. Fortunately, the crew was able to get the victim's history from bystanders and managed to keep him breathing until the paramedics arrived.

According to Moy, the emergency dramas in which he takes part are seldom as satisfying as they are on TV dramas. "Each of us plays a little part. We keep someone alive until he or she can be transported to the hospital, and then the emergency room staff takes over. We have to do historical research to discover a victim's status, and usually never learn the final outcome."

Cambridge fire fighters can be reassigned to fill job openings at other stations in the city. Once a fire fighter's probationary year is completed, he or she can put in a request for a transfer. Arthur Moy imagines that someday he'd like to move on to other Cambridge fire companies and stations—perhaps even to the Cambridge rescue units—to gain experience and become a more well-rounded fire fighter. These units are well respected throughout the state and, even within the department, are considered the elite. Their job is to get to a fire and perform initial search of the structure, often entering ahead of the water attack lines. The rescue units also perform hazmat operations and confined space and high-angle rescues.

"Each of us plays a little part [in emergency dramas]. We keep someone alive until he or she can be transported to the hospital, and then the emergency room staff takes over."

For the time being, Moy has not requested transfer from Engine 9, staying loyal to his group and crew until department needs dictate otherwise or until a sweet opportunity knocks. "When we got out of the academy, they told us to expect to see less action on the job," he says. "and they were right. In a way, it was a letdown, not unlike like soldiers without a war." They have to stay vigilant against complacency, explains Moy, so he and his fellow fire fighters train and retrain their procedures. They try to anticipate future situations, needs, and procedures.

"We fire fighters want more fires to extinguish, but, of course, we don't want people's lives and property to be placed in jeopardy. That's why the academy experience was so good. You got to fight fires without risking people's lives or property."

Mentors

9

Joe Murabito

Joe Murabito

PROFILE

Fire Department: Delaware State Fire School

Location: Dover, Delaware

Position / Title: Director

Rookie Year: 1966

Height / Weight: 5′10″/ 195 lbs.

Previous Job: Wholesale and retail, lumber industry

Hobbies / Other Interests: Woodworking, sport shooting

Education / Training: AAS Emergency Services Management, Certified Instructor III

Reason for Joining the Fire Service: At first, the appeal of the fire department was primarily adventure. Since childhood, Murabito had been excited at the sound of fire sirens. As a boy, he'd chased the fire trucks on his bike to get to the action.

Most Memorable Rookie Experience: Halloween night fire on the second floor of a two-story house, with children trapped inside (as recounted here).

Fire Service Ambition: To be a fire chief and work for Delaware State Fire School

Joe Murabito was on the first truck out of the Lewes, Delaware, firehouse at 4:30 A.M. on Halloween night in 1968. In those days in Lewes, fire alarms were called in to the local "power house," the electric-generator station, where whoever was on duty would trip a remote switch at the firehouse to sound the siren. If you were in the firehouse when this happened, you knew the siren was about to blow, because the relay made a loud whamming sound. The power station would then contact the firehouse by a direct phone line to give the details, and all available volunteers would report to the station, never directly to the fire scene. The call relayed to the all-volunteer department that night was to a fire on the second floor of a two-story house. Children were trapped inside.

As the truck Murabito was riding arrived at the scene, one child had already jumped from an upstairs window and had landed with his clothing on fire and a second one jumped just after the arrival of the engine. Murabito smothered the flames then went up a ladder—protected by only a cotton coat, a fiberglass helmet, and 3/4-length boots—to rescue a second child trapped at a window. But by the time Murabito reached the window, the child had disappeared.

"It was traumatic," Murabito recalls, "seeing the kid's face when I started up the ladder, and then it wasn't there when I got to the top." The child died in the blaze. The second child to jump died in the hospital a week later. Murabito suffered burns on his hands and face, which were bandaged at the local hospital.

In 1968, with America embroiled in the Vietnam War and with urban riots and political assassination the facts of life, the prevailing attitude toward witnessing death and destruction was unsympathetic at best. "That incident bothered me at the time, but it bothered me more later when I had my own kids," says Murabito. He remembers having to explain his injuries the next day to his classmates at the community college where he was taking courses. There was no question of time off to recover from the shock, no mandatory counseling. "You were supposed to suck it up," says Murabito. "There was no stress management. When I think back on it, I don't think I'd want to subject kids to that today."

Family and Background

The night of that tragic call, Murabito had been in the fire service less than two years and was still considered a junior member. Four years later, he would become the youngest fire chief in Delaware history when he took charge of the Lewes company where he'd received his hands-on training.

"[In 1968] there was no stress management [after a tragic incident]. When I think back on it, I don't think I'd want to subject kids to that today."

A career trajectory that steep would be unheard of today, especially in a populous industrialized state such as Delaware. But in his first years, Murabito, who is now in charge of Delaware's Dover-based statewide fire service training program, was facing the demands and dangers of fire fighting with little formal training. Field promotions were common and, for the most part, recruits learned by doing. At that time the only formal training was a yearly two-weekend course offered in cooperation with the newly formed fire school and the State Firemen's Association.

Murabito joined the Lewes brigade, at the encouragement of a friend, at age 17, while he was still in high school. Because he was so young, Murabito needed permission to sign up from his father, a first generation Italian-American. His father, who was one of ten children, was a schoolteacher with a master's degree in education. Joe was the first (and, so far, only) member of his family to work in the fire service. "I'll sign it," said his father, "but you won't last six weeks. It's dirty, nasty work." Decades later, Murabito and his father, now in his 80s, still laugh about the inaccuracy of that prediction.

At first, the appeal of the fire department was primarily adventure. Since childhood, Murabito had been excited at the sound of fire sirens. As a boy, he'd chased the fire trucks on his bike to get to the action. Thanks to his first mentors, he would soon develop an appreciation for the real workload and time commitment that went along with the excitement.

Lewes, a beach resort on Cape Henlopen in southeast Delaware, has experienced dramatic growth as a tourist spot since Murabito's high school days. In 1966 the Lewes volunteer fire service, one of 60 volunteer houses in a state with only one paid department (in Wilmington), answered 75 fire calls and 300 ambulance calls. Today, the department fields 500 fire calls and 2000 ambulance calls annually. However, like many departments, Murabito notes, the department may actually see fewer fires because there are more joint responses and more calls from faulty alarms.

In the 1960s there was little formal training. Field promotions were common and, for the most part, recruits learned by doing.

The 1966 Lewes Fire Department had 45 members and one firehouse located in what's now very expensive real estate in the center of town. Volunteers were required to live within a mile of the town limits. Today, Lewes is protected by 100 volunteers, working out of three stations strategically located to cover the entire fire district.

Compared to the rigorous training and probation systems in place at even small fire departments today, the loose organization that greeted Murabito in Lewes may seem lax. But in the late 1960s, the Lewes Fire Department was considered progressive, because it was staffed, in part, by four career fire fighters from Dover Air Force Base, roughly 45 miles up the Delaware coast. Thanks to their military experience these veterans, Murabito explains, understood the value of formal instruction to supplement what the young volunteers learned on the job. Lewes also had what most departments lacked: a junior members' program for beginners, which accepted recruits as young as 16. This was the program that brought Murabito to his lifelong career in the fire service. He held the status of junior member with the Lewes brigade until he was 21 years old.

When Murabito joined up, in February of 1966, four months before his high school graduation, he assumed that belonging to a volunteer fire department meant only being available for calls. "But the Air Force guys," he remembers, "expected discipline. They told me I had to be committed to the firehouse and that I had to 'be there.' " And he was there—a good deal of the time. Most nights, Murabito would come home from school, do his homework, and then report to the fire station. He also gave up a lot of weekend free time.

"The Air Force guys expected discipline. They told me I had to be committed to the firehouse and that I had to 'be there.' "

"Once, they called me and told me to report on Saturday morning to help wash the trucks. I was a kid. I told them I had something else to do, and they let me know. They said, 'You have to do it if you want to be in the fire service.' "

If a Lewes junior volunteer was in high school, he had to stay on track at school and at home as well. Young rookies were required to show their report cards to the fire chief. If a high school rookie's grades fell below a B average, the chief put him on probation until the next marking period. The system worked, Murabito remembers. "Kids went on probation, but nobody dropped out. They went back and got their grades up."

There were also special rules that applied to junior volunteers—rules designed in part to keep the fire service from disrupting the rookies' educations and in part to mollify parents who may have had reservations about teenagers being on twenty-four-hour call. Juniors weren't allowed to respond to a call after 10:00 P.M. or before 6:00 A.M., for example, except on weekends and holidays. Murabito remembers sitting in the firehouse on school nights, hoping that if there was an alarm, it would be called in before 10:00 P.M. because once a junior volunteer was on a late-night call, he could stay.

Murabito's early training was informal but far from casual. Those old hands could be relentless taskmasters.

Training

In addition to assigning him the routine chores that traditionally fall to rookies, the old hands at the Lewes station drilled newcomers in basic fire-fighting techniques. Murabito's early training was informal but far from casual. Those old hands could be relentless taskmasters, he recalls. One week, they'd give a rookie a tour of a fire truck, defining and explaining all the parts in detail; the next week, the junior member would be expected to recite those descriptions back to them.

There were also opportunities for learning that today's rookies simply don't have. The junior members kept the apparatus clean and did a lot of the gofer work; but in Delaware in the 1960s you didn't need credentials to ride a truck, so the old-timers let the beginners go to fires. Amazingly, Murabito went on his first response within his first week in the department.

The junior members kept the apparatus clean and did a lot of the gofer work; but in Delaware in the 1960s you didn't need credentials to ride a truck, so the old-timers let the beginners go to fires.

It was a 5:00 P.M. fire in a low-income neighborhood. Murabito rode the first truck with three veteran volunteers. The fire, it turned out, was in a kitchen stove. Murabito, totally inexperienced, went in as backup behind one of the vets, humping hose. Neither wore any breathing apparatus. The older man worked the nozzle and extinguished the fire quickly. Murabito emerged coughing but exhilarated. In his first year, Murabito estimates, he went on 90 percent of the department's 75 fire calls.

The fact that newcomers went on fire calls sometimes placed a lot of responsibility in very young and inexperienced hands. Murabito recalls having a conversation early in his career with one of the company's more experienced fire fighters who'd been to fire school. The veteran told Murabito that in school they'd taught the students to open the nozzle, stick it in the door of a burning room, and just whip it around. The technique didn't sound very sophisticated or professional. That same night, Murabito found himself holding the nozzle end of a fire hose behind a three-story house with flames blowing out onto the back porch. Lacking any better idea, he did what he'd been told. To his astonishment, the fire went out.

Murabito's rookie training, however, wasn't all by the seat of his pants. The Air Force fire fighters in the Lewes department, Murabito's first mentors, insisted that rookies receive as much formal training as the state provided. At the time, courses were given over only two weekends twice a year. By April of his rookie year, Murabito was in fire school. The curriculum was basic. The first round of training consisted of lessons in ladder climbing and using canister or chemox masks. The second set of classes covered fighting a kerosene fire in a dirt pit and using fire extinguishers. Still, the training experience provided an incentive to improve. At classes and back at the Lewes station, there was a spirit of competition among the rookies. Junior members asserted mutual peer pressure to be good fire fighters, and the older rookies made a point of bringing the newcomers up to speed.

Meanwhile, Murabito began to realize the social benefits of being a high school student on the town's volunteer fire department. In a small community such as Lewes, everyone knew when there'd been a fire, and Murabito became something of a teenage celebrity—routinely asked by classmates and neighbors to recount the details of the previous day's alarm.

At school, his principal put together a student "Fire Patrol" to assist with and monitor fire drills. All the students who served with the volunteer fire department were appointed, and Murabito soon became co-captain. Since his fellow students knew that he actually fought fires, he got a lot more respect than the average hall monitor did.

"It was all recognition from my community," Murabito explains, "and I liked it. There was also a macho-ego aspect to being with the department that made me feel important and part of something."

At the Lewes department, though, there was no elaborate hazing. Informal initiation was limited to silly pranks such as filling a rookie's boots with packets of ketchup from MacDonald's. For the most part, Murabito recalls, "You became a member of the team by working hard and making contributions to the department—whether polishing the apparatus or working a fire."

"You became a member of the team by working hard and making contributions to the department—whether polishing the apparatus or working a fire."

Recognition at home and at school fed Murabito's ego, but he came to realize the real and lasting gratification of being in the fire service when he went on his first ambulance call. Murabito was a lifeguard in high school and had had extensive first-aid training. He became a first-aid instructor, teaching classes to the local Girl Scout troop and other community organizations. Because of this experience, Murabito was eligible to go on ambulance calls during his rookie year.

The Lewes Fire Department was not well equipped. Its rescue vehicle was a panel truck the department had bought from a neighboring community who'd upgraded to a larger, more professional vehicle. The truck was outfitted with wooden shelves that left a narrow aisle for passengers. There was so little equipment that, 36 years later, Murabito can still recite the list of gear. The department also had an ambulance that responded to most of the nonfire emergency calls.

As soon as it was suggested, Murabito was eager to assume ambulance duties. He remembers having a boyish attraction to the uniform. Ambulance volunteers wore white coveralls with the name of the fire department embroidered in red on the back. "It made you feel important," he says. He couldn't wait to put the suit on and go on a call. The small-time glamour of his ambulance duties was, however, quickly overshadowed by a more serious reality.

Murabito's first ambulance call was to a car crash that involved the local IGA grocer and his wife. Suddenly, at age 17, Murabito found himself helping people he'd known most of his life, and he realized that it felt good. It was satisfying to play a responsible role in what were sometimes life-and-death situations. "I realized that I liked helping people," he says today. "It was satisfying, and you have to get satisfaction out of the fire service if you're going to last."

Joe Murabito served his rookie days with the Lewes Fire Department during the Vietnam draft, when young men were conscripted at

"I realized that I liked helping people. It was satisfying, and you have to get satisfaction out of the fire service if you're going to last."

age 18. To avoid being drafted into the infantry, Murabito got his parents' permission to sign up for the Air Force. He took his preinduction tests when he was 17, but the Air Force put him on hold because it gave priority to the 18-years-olds who were threatened with the Army draft. When Murabito reached draft age, he still hadn't heard anything, so he called and discovered that the recruiting office had lost his paperwork. He quickly enlisted again, but within a day he was classified permanently exempt (4F) because his medical tests uncovered symptoms of a potential condition that made him a high statistical risk to end up on military disability. It was, the Air Force explained, more cost-effective to send him home than to take the chance of having to support him on disability, so Murabito found himself back in the Lewes fire station the next day. His health has been fine ever since.

Career Development

With military service out of the picture, Murabito was free to concentrate on a civilian career. It's here that two other of Murabito's mentors stepped in. The Delaware State Fire School was growing, as the result of a campaign to offer a broader selection of courses more frequently. Although many of the state's departments required volunteers to take state-sponsored training, that requirement often translated into fire fighters taking the same courses year after year. That was about to change.

In 1967, long-time director of the Maryland Fire and Rescue Institute, John Hoglund, came to the school to run officer training courses. Murabito signed up for one of Hoglund's classes. During one session, Hoglund held a taxing mock-crisis exercise. He spread a map of a fire district out on the table and handed out toy fire trucks to the students. The object was for the students to deploy their resources as effectively as possible, as Hoglund piled on one hypothetical emergency after another. Hoglund's goal was to overload the situation to the point where the students would throw up their hands.

The example that was designed to break the back of the imaginary fire department's resources involved a fire in a church. Hoglund explained that the church was an old and valuable property, beloved by the community and deeply rooted in the town's history and tradition. Murabito got Hoglund's attention when he suggested letting the church burn because it was empty and isolated, and the trucks were needed elsewhere. Hoglund was impressed with the seemingly ruthless approach that, in real-life circumstances, would have saved the most lives. Murabito was acquiring a reputation at the fire school.

The first faculty member to hold the position of senior instructor at the school was author and lecturer Harvey Grant. Grant, who wrote a well-known series of emergency-care texts and has been called "the father of vehicle extraction and rescue," (the International Association of Fire Chiefs still gives the Harvey Grant Excellence in Rescue Award) was a firm believer in attracting young people into the fire service. Grant asked Murabito, then still 18 years old, whether he would like to join the school's cadet instructor program. Murabito jumped at the chance. He took an education methodology class one night a week and signed on as a junior instructor. The instructors in the cadet program, he soon learned, mostly did the school's gofer work; but, just as Grant had expected, it was the beginning of Murabito's long and successful association with fire training.

However, neither part-time teaching nor serving with the volunteer fire department constituted a career. In 1969 Murabito attempted to join the paid fire department in Wilmington. He took the tests and did well, but ended up twenty-eighth on a waiting list of twenty-nine recruits. The examiner suggested that he join the police force, which had immediate openings, so Murabito served as a Wilmington police officer for more than two years. During that stint, he maintained his relationship with the Lewes Fire Department by going home on weekends and going out on calls.

Murabito left the police force and did some retail work at a lumberyard while still teaching part-time at the fire school. In the early 1970s, he was appointed chief of the Lewes Volunteer Fire Department, and was hired at the fire school in 1978 as a senior instructor to oversee programs and run weekend classes. In 1989 he was appointed the school's assistant director; and in 1997 he became its director, replacing another of his long-time mentors, Lou Amabili. Amabili,

Murabito credits the America Burning *report, which described the outmoded state of America's fire-fighting capabilities, with modernizing the fire service in the United States.*

Delaware fire departments are independent corporations with bylaws that require rookies serve a two-year probation period and take classes in basic fire fighting, structural fire fighting, hazardous materials handling, vehicle extraction, and more.

who had held the position for thirty-two years, had served on the 1972 presidential commission that produced the well-known *America Burning* report. Murabito credits this report, which described the outmoded state of America's fire-fighting capabilities, with modernizing the fire service in the United States.

Today's Fire Service in Delaware

As Amabili cautioned when he handed over the job to Murabito, the fire services in Delaware today are very different from what they were when he took over the fire school's reins. Delaware fire departments are independent corporations with bylaws that require rookies, in many cases, to serve a two-year probation period and take classes in basic fire fighting, structural fire fighting, hazardous materials handling, vehicle extraction, and more. Some departments have specific classroom requirements for individuals to qualify as officers. The Delaware State Fire School now runs classes every weekend of the year, except on holidays and during the months of July and August.

Throughout the state, the emphasis on training has grown substantially since Joe Murabito worked his first fire. The fire school, which has a 2 million-dollar annual budget, has grown to meet the need. Each weekend, several hundred rookies and seasoned fire fighters converge on the Delaware State Fire School in Dover, where Murabito now lives. They come for routine as well as specialized training. The school provides a variety of training—from industrial fire training to *Risk Watch*™ and other fire safety programs—as well as overseeing all ambulances operating in the state and certifying EMTs. The school employs 8 full-time fire- and rescue-training administrators and uses 300 part-time contract teachers, each a specialist. In 2001, the Delaware State Fire School offered 110,000

The Delaware State Fire School in Dover provides a variety of training—from industrial fire training to Risk Watch™ and other fire safety programs—as well as overseeing all ambulances operating in the state and certifying EMTs.

student hours of training to 9000 students, including specialized training in specific areas, such as the use of breathing apparatus, and courses for drivers and rescue teams.

Now it's Murabito's turn to be a mentor to the next generation of fire fighters, and he delights in the role. After all, he owes his own career to the veterans who took an interest in rookies.

10

Minor Misadventures

Kyle Page

Kyle Page

PROFILE

Fire Department: Carrollton, Texas, Fire Department

Location: Carrollton, Texas

Position / Title: Fire fighter / EMT

Rookie Year: 2000–2001

Height / Weight: 5′8″ / 220 lbs.

Previous Job: Call taker at collections center

Hobbies / Other Interests: Sports, movies, live music, volunteer work

Education / Training: Working toward a degree in emergency management administration, University of North Texas

Reason for Joining the Fire Service: Wanted a job with variety and excitement and that would help people

Most Memorable Rookie Experience: Destroying a light panel at the psychiatrist's office when he went for his polygraph

Fire Service Ambition: Page wants to end every shift knowing that he did everything in his power to provide the extraordinary level of care—for his patients and for their property—that is expected of his department. He wants to know that he made a positive difference for someone.

Kyle Page missed what should have been his first fire. He'd just finished his brief break-in period—riding along as an observer for several shifts—at the Carrollton, Texas, Fire Department and was about to begin on-the-job training as a full-fledged rookie. As fate would have it, though, he spent his first shift driving 30 miles to Arlington to take the state EMT test, and while he was gone, a call came in for a fire at an auto-parts manufacturer. He was sorry he missed the call. It was the kind of fire the 26-year-old Page describes as "just big enough to have fun on."

Besides, even a medium-sized fire call is rare in Carrollton, a 38-square-mile city of 115,000 northwest of Dallas. Suburban, middle class, and generally affluent, Carrollton offers a brand of fire fighting that's a far cry from what goes on in nearby Dallas. Crime in Carrollton is low, and the majority of housing is new and well protected from fire. So Page spent much of his probationary period at Carrollton longing for some excitement.

Family and Background

Page's interest in joining a fire department had its roots in his childhood in a remote and rural community with the unlikely name of Cool, Texas, where his grandfather was a volunteer in the local fire brigade. The town of Cool maintained a small and decidedly informal volunteer department. "For a while, the apparatus were parked in our front yard," Page recalls.

Since he had no training or experience, however, Page never joined the Cool volunteers. As he went through school he lost track of his early interest in fire fighting. Page enrolled in a junior college not too far from home, then moved on to the University of Texas at Arlington, where he "changed majors a lot. I had a sporadic educational career," he says. "I was in and out of several different colleges." He then attended a local community college and is currently working toward a degree in Emergency Management Administration back at the University of North Texas.

What really pushed Page to apply to be a Carrollton fire fighter was his hatred of office work. At one point in his early career, Page worked in a call center making collection calls. "It left a bad taste in my mouth," he reports. "I decided I never wanted to work in an office again. I prefer to be outdoors anyway, but that job left me with the idea that if you work in a business, you have to be willing to take advantage of people to make money for your company. That just doesn't sit well with my personality."

"I decided I never wanted to work in an office again."

Around the time Page was unhappily toiling at the call center, his stepbrother was trying hard—but with little success—to land a job with the Fort Worth Fire Department. His stepbrother's attempt renewed Page's interest in his grandfather's avocation and inspired him to investigate a career in fire fighting.

Testing and Hiring Process

Carrollton follows a typical application and screening process, common to most of the Texas state civil service departments. Page applied and was told to show up for the initial aptitude test, which made him nervous, he says, because there were no study materials. Much of the test focused on general aptitude—reading comprehension and math skills—but it didn't stop there. The test included map-reading questions, such as finding the shortest route from a given location to the hospital, for example, as well as "situational" questions tailored to would-be fire fighters. "They asked questions like, 'Say you're working at a fire scene, severely undermanned, and a man in civilian clothes who might be drunk offers to help. What do you do?' " recalls Page.

Page passed the written exam and was among the top twenty-five scorers who were assigned to take the physical ability test for Carrollton. That test was slated as a two-weekend process. Recruits had the option of participating in an orientation session the weekend prior to the actual test; then, if they weren't scared off by the requirements, they came back the next weekend for the test itself. Page was on two weeks of active duty for the Marine Corps Reserves the weekend of the orientation, so he showed up on the weekend of the test not knowing what to expect beyond what he had heard in casual conversation.

It was a difficult test. In one of the tasks, recruits were required to climb a 75-foot aerial ladder while wearing a 50-pound vest. "We had two minutes to do that," remembers Page. "Anyone could have done that. I think that was mostly to test if we were afraid of heights." Other, more difficult tasks, however, included carrying a high-rise fire-

"It wasn't easy, but it boggled my mind that some people didn't finish. I don't know why anybody would quit. I think it all has to do with mind-set."

fighting pack up four flights, hooking up a 3-inch hose, advancing a charged (water-filled) hose line 75 feet, crawling across ceiling joists for 15 to 20 feet, dragging three sections of dry 3-inch hose 50 feet and back, and dragging a 150-pound dummy 50 to 70 feet.

"It wasn't easy," reports Page, "but it boggled my mind that some people didn't finish. I don't know why anybody would quit. I think it all has to do with mind-set."

The test was pass / fail. Page passed, which moved him along to the background check stage of the application process. "I had no idea they were going to ask such in-depth questions," Page says. "They asked about my employers all the way back to when I was sixteen. I spent a full day looking up all the answers."

The next two steps of the application process—an interview with a board of members of the Carrollton Fire Department and a polygraph test—turned into more misadventures for Page, that, fortunately, he can laugh about after the fact.

"I was nervous about the interview anyway. I showed up wearing a polo shirt and slacks and then I saw that the other candidates were all wearing suits. I thought I was done, right there." But he wasn't, and the interview went well despite Page's informal attire. The polygraph test was another story.

Page showed up early at the office of the psychologist who was administering the polygraph and took a seat in the waiting room with other fire-fighting candidates from Carrollton. A wasp had flown into the room and was hovering around the fluorescent lights in the ceiling. Page, who has a minor allergy to wasp stings, asked the psychologist's receptionist if she minded if he killed the insect.

"She said okay," says Page, "so I grabbed a magazine and took a swing and knocked out about a quarter of the light panel!"

Despite his dramatic entrance, Page passed the polygraph and, soon after, he passed the additional psychological and physical examinations. A week later, Page had a firm offer of employment from the Carrollton Fire Department and was told to report for orientation in October of 2000. He was both surprised and grateful because he'd known other people who'd been trying to land fire-fighting jobs for three to five years.

Training

At Carrollton, a fire fighter's rookie year begins when he or she enters academy instruction, which is taken at the Dallas fire-fighting academy; most rookies, therefore, spend only three or four months on probation doing live shift work.

Because the Carrollton Fire Department was low on staff in 2000, Page was one of eight candidates hired. Of those eight, two were already certified fire fighters and entered the department by running five orientation shifts. Page and the remaining five set off for the Dallas fire academy accompanied by a captain from the Carrollton department to help with their instruction.

Page had friends with the Dallas Fire Department and had heard that the academy had a militaristic environment—similar to that of a boot camp. He wondered what to expect. Would it be like Marine boot camp or would it be something of a less military nature?

What he found when he arrived was what he tactfully describes as a "very structured environment"—one that was military in style, but not as severe as what he'd imagined. As Page and his fellow recruits arrived, they met the previous class on its way out, members of which were quick to assure the newcomers that they'd never survive. "They were very encouraging," says Page, sarcastically.

Page's first day at the academy seemed easy enough to survive. It consisted of a series of orientation lectures and sessions covering basic human resources topics, such as sexual harassment policies and advancement. The first surprise came at lunchtime. Just as Page and his classmates were about to be dismissed for lunch, they were ordered to fall out and do a lap around the academy's one-mile track.

After that, they did get some lunch and returned to more classes. That lap, however, was part of a trap. After class sessions, the unsuspecting group was told to fall out again and report to the track. Except that this time recruits from the academy's previous graduating class were waiting for them with a full array of training apparatus. As the newcomers circled the track, they were blasted with streams of water and splattered with mud as the former students sprayed hose streams into puddles beside the track.

Having survived that initiation, Page and his fellow recruits were next harangued by a Dallas Fire Department lieutenant about their place in the academy pecking order. "He was as salty as they come," remembers Page. "He talked to us like the drill instructor in *Full Metal Jacket.*"

The outgoing class also left another souvenir for the "newbies." They switched the salt and the sugar at the captain's table—a prank for which, of course, the new class was blamed.

The academy's protocols were indeed military, which, thanks to his experience in the reserves, wasn't totally foreign to Page. Two weeks into their academy training, however, Page and several other recruits from Carrollton became an unpleasant object lesson in discipline.

"It was much like in the military," Page explains. "If an officer walked in the room, you were supposed to come to attention—immediately. A bunch of the guys—mostly from Carrollton—were sitting around studying together. We had our noses in our books when the captain from Carrollton who had accompanied us to the academy walked in. He must have taken four steps before somebody noticed him and called us to attention. The captain said, 'That wasn't fast enough. You owe me.' "

What the Carrollton recruits owed their captain became clear at lunch a few days later. In a drill designed to simulate the real-life interruption of a fire call, academy officers rang the fire bell during lunch hour and the class was expected to rush to their lockers and report to the training field for inspection in full gear. The alarm rang and the recruits scrambled.

"I ended up standing in the ranks next to a guy whose dad was a chief in the Dallas department," Page recalls with a laugh. "I figured he should have known what he was doing, but he turned out with his helmet on backwards. It was an understandable mistake, as we had received no instruction on donning our gear at that point."

"We did get there fast, but not fast enough. The section chief let us know that and made us do push-ups. Then the captain said, 'Where are those bodies that owe me?' " Page and the other recruits who'd ignored their captain raised their hands and were rewarded with more push-ups. To drive the point home, the rest of the class was then told to run laps on their behalf.

"One of the guys cooked his specialty—Mexican food—and it was a great meal, so we gorged ourselves. That must have been what the instructors were waiting for because as soon as we finished eating, they had us run the track."

Lunchtime, it turned out, became a routine opportunity for unwelcome training drills. "We had been warned that the instructors would have us run the track right after we ate," says Page, "but three months into the academy they hadn't done it, and we didn't believe it would really happen. The food was not all that great. In fact, some meals would choke a maggot, so we never ate much at lunch."

"But one day, one of the guys cooked his specialty—Mexican food—and it was a great meal, so we gorged ourselves. That must have been what the instructors were waiting for because as soon as we finished eating, they had us run the track." Needless to say, a lot of the recruits lost their lunches.

Despite the training rigors and military excesses, only one member of Page's class tried to drop out of the academy. At 36, he was older than most recruits and the physical training took more of a toll on him than the rest of the class. He was discouraged by the physical strain of training and decided to quit, but his classmates wouldn't let him.

"We inspired him to stay," says Page. "We took care of each other like that."

Some of the academy's training drills were extremely taxing. Every other day, the recruits trained with what was called the "red bag" ("even though the bag was really green," notes Page). The red bag was a sack loaded with cut-up sections of 5-inch fire hose. It was incredibly heavy, and recruits were told to carry it up and down a 6-story tower three times. At first, cadets did the drill in their gym clothes, but eventually they worked up to carrying the red bag while wearing full gear.

"We inspired [an older recruit who wanted to quit] to stay. We took care of each other like that."

Another tough exercise involved dragging a length of fire hose attached to a fire truck tire. It was during this exercise that the 36-year-old tore a calf muscle and was forced to drop from the recruit class after all. To his credit, he resumed training with the following recruit class and did very well. He now works for Dallas Fire and Rescue as a fire fighter / paramedic.

After six months of fire training, Page and his class were ready to graduate. Graduation took place during a dramatic ceremony known as "Burn Night." The recruits' families were invited to attend and watch the class give a rappelling demonstration, put out a live fire, and confront a simulated residential propane tank fire.

After the fire academy, the recruit class went on to EMT school at a community college program run in conjunction with the University of Texas Southwestern Medical School. Some members of Page's recruit class stayed on to complete their paramedic training at the Dallas facility, but the Carrollton recruits were shipped back to Carrollton to begin their orientation shifts.

The Carrollton department's paramedic staff is capped at 75, so not all Carrollton fire fighters are required to hold paramedic certification. If the department is fully staffed with paramedics, a Carrollton fire fighter's only option is to pay for his or her own paramedic schooling or submit a written request to the department and wait to be trained as paramedic positions open up. All recruits follow up their fire training with a month of EMT school before entering a three- to five-shift break-in period with the Carrollton department. It was this arrangement that caused Page to miss his first-shift's fire because he was off in Arlington testing for state EMT certification.

Rookie Year Experience

Page's first real fire call came when he'd been running shifts as a rookie for two months. He was out on a truck doing pre-fire surveys—that is, visiting businesses to inspect their fire protection systems and learn the layout of their facilities. The crew was gassing up the truck on their way back to the station for dinner when they were called to a fire nearby. Page's was the second truck to arrive at the scene and there wasn't a lot left to do. Still, Page was thrilled. He was "very excited to get a real fire," he remembers, even though his job ended up just going in and pulling down the building's ceiling.

Because the volume of fire calls is low in Carrollton, fire fighters change station assignments every three to six months. This system keeps the younger fire fighters from growing stale as a result of being assigned to a slow station too long.

Page was eager for some serious fire-fighting action. He considered himself lucky when his truck showed up first at a garage fire and he got to go in with the nozzle. Then, on a hot and humid day in August, shortly after his probationary period ended, Page got his first taste of major league fire fighting. Initially, the call came in for an investigation. Then, as Page waited at his station, it was bumped up to "full box" (single-alarm fire), then to two alarms.

"I learned that all a big fire means is humping around a lot of heavy hoses. It was tough work."

The fire was in a three-unit condo with a flat wood-shingle roof. The building was heavily engaged. The call was Page's chance to fulfill all his heroic notions about fire fighting. It was also, as it turned out, an opportunity for a first-hand lesson in the realities of the job.

"Fire fighting is the best job in the world."

"At first, I was having a good old time," Page recalls. "Then real quick, I was not happy. I learned that all a big fire means is humping around a lot of heavy hoses. It was tough work." The experience didn't dampen Page's enthusiasm for the fire service; it just put things into perspective.

Page's probationary months with the Carrollton department were fairly comfortable. There was some stress, especially feeling under pressure not to do anything stupid, he recalls. Page also had to learn to seek out advice. "The senior guys weren't going to hold your hand and walk you through everything," he says. "but if you expressed interest, if you asked questions, they'd teach you everything they knew."

"Sometimes, when you're on probation, the guys mess with you, good-naturedly . . . kinda," Page continues. "People gave me a hard time—treating me like a goon. Everybody tells you that's a good thing, that it means they like you. And after you've been there a while, you realize that they do ignore the guys who aren't well liked."

Now that he's off probation, Page looks back at his path to the fire service as a time of hard work and minor misadventures—from the wasp in the psychologist's office to missing his first chance at a real fire. And he's not unhappy that it's over. Echoing the sentiments of so many young fire fighters, Page contends that "fire fighting is the best job in the world."

As a full-time fire fighter, two things have surprised him. The first was the adjustment to shift work (24 hours on, followed by 48 hours off). This unusual schedule is difficult to adjust to initially. "One time, I was driving somewhere with my girlfriend and I was hit by a wave of anxiety. I suddenly panicked that I was supposed to be at the station."

The other is how much he enjoys the job. "I've never had a job where you wake up and you're tired or sick but you go to work just because you love the job so much," he says.

"I've never had a job where you wake up and you're tired or sick but you go to work just because you love the job so much."

"I'm amazed at how many people I know express interest in fire fighting but never do anything about it. I tell them they have to join for the right reasons. It's not about the downtime or the two days off when you can have another part-time job. And it's not about the time just hanging around the station, waiting. If that's why you want to join, you don't belong."

Building Mutual Trust

11

Jennifer Steele

Jennifer Steele

PROFILE

Fire Department: Kitsap County Fire District 7

Location: Port Orchard, Washington

Position / Title: Fire fighter

Rookie Year: 2002

Height / Weight: 5′4″ / 150 lbs.

Previous Job: Student, personal training

Hobbies / Other Interests: Running, archery, violin

Education / Training: A.S. in fire fighting, Olympic Community College; Kitsap County Volunteer Residents Program

Reason for Joining the Fire Service: Wanted a career that allowed her to help others and a job that was constantly different and challenging

Most Memorable Rookie Experience: "All the time I was driving [a pumper on a garage fire call], I was nervous because I knew I'd have to pump. I kept thinking about pump-discharge procedure—going over everything I'd learned in class. I just did what I'd been trained to do [and didn't] make any major mistakes. It felt tremendous."

Fire Service Ambition: "I'm looking forward to staying with my current department and continuing to build relationships with coworkers and the community. I would also like to develop and build on my skills and abilities and maybe eventually work into a leadership role."

By now, Jennifer Steele has responded to follow-up calls so often that she's on a first-name basis with the diabetic man she treated on her first medical emergency run. She remembers the first time, however, riding in the back seat of the medical unit, with its sirens blaring, and thinking to herself, "What am I going to do?" At that point in her fire service career, Steele's emergency medical system (EMS) skills were minimal; she didn't even know the medic with whom she was working; and she was preoccupied with doubts that she had anything to offer. At the scene, she felt useless. She fumbled. She wasn't familiar with the rig and had trouble fetching gear. She was distracted by the sight of a man bordering on a diabetic coma. Her team was patient with her. When they got back to the station, they explained what they'd been doing at the scene, and why, and what they needed and expected from her on calls.

Today, Steele is sometimes amazed at the casual attitude inside an ambulance as it rushes through traffic to an emergency. As cars pull off the road to make way, the driver and crew might be chatting about sports or dinner plans, confident that when they get to the scene they'll be prepared to work as a team. "Each time you succeed, your confidence goes up," Steele explains. "You know the people you're working with and you trust each other. Your on-the-job relationships are important because you have to have confidence in the abilities of your coworkers."

Volunteer Residents Program

Technically, Steele wasn't even a rookie with Washington state's Kitsap County Fire Department when she went on that first medical call. She was a volunteer in the district's Volunteer Residents Program, a system that offers novices access to the fire service while increasing the departments' staffing beyond its budget's ability to hire full-time fire fighters.

"Your on-the-job relationships are important because you have to have confidence in the abilities of your coworkers."

Under the Volunteer Residents Program, trained volunteers live in the firehouse—some permanently, some on rotation—and pull shifts, usually overnight. Steele is married to a Navy nuclear engineer, currently in officer's training at the University of Washington in nearby Seattle, so she arranged to volunteer every third day, checking in at 5:30 P.M. and staying through to 8:00 A.M. Residents sleep in the station's bunks, respond to calls, and even drive the first-aid car and medical unit on calls. In exchange, volunteers are offered a $300-per-month stipend or funding for education. Steele took advantage of the latter opportunity to study for an associate degree in fire science, a program that includes Fire Fighter I training, as well as emergency medical technician (EMT) and hazardous material (hazmat) certification, at Olympic Community College in nearby Bremerton, Washington.

She was also able to parlay the resident experience into a position as full-time fire fighter. Being a resident is invaluable hands-on preparation for the fire service, and, because the program offers a foot in the door to the Kitsap County Fire Department, competition to become a resident is high. When Steele applied to be a volunteer resident, there were 80 candidates countywide, 30 of whom were eventually accepted. Joining the residents program involves roughly the same process as being hired as a full-time fire fighter. Steele was interviewed and underwent both physical and written exams. Once she was accepted, she spent five weeks at a training center—"a typical militaristic boot camp," she recalls—before she was qualified to go on calls.

The Kitsap County Fire Department gives its residents a lot of responsibility. For example, they're required to learn to drive all the department's vehicles, including the trucks. "I'm a good driver," Steele says, "but the trucks are big. I embarrassed myself a few times. Drove over some curbs or had to back up to make a turn. I can do it now. It made me a better car driver." Because Steele didn't have a full-time day job, she was sometimes able to pull 24-hour shifts as a resident. That gave her the opportunity to work days, when there are typically more alarms, and to get to know more of her colleagues. Those friendships and the added exposure to experienced fire fighters, she says, gave her a big advantage and prepared her for the rigors of the job of fire fighting.

Family and Background

Jennifer Steele never expected to leave her home state, never mind becoming a fire fighter. Steele was born and raised in a close-knit family in Palmer, Alaska, not far from Anchorage, the third of four children. She has two older brothers and a younger sister. She met her future husband at age 14 or 15 while they were in high school and was married at 19. Steele, whose interests range from playing the violin to archery, originally wanted to be a writer and was enrolled in a state college journalism program when her husband decided to join the Navy. Overnight, the couple became a Navy family. Steele gave up scholarships in Alaska and went to Orlando, Florida, during her husband's Navy training. While there, she had a change of heart and, deciding she'd prefer a more physical career than reporting would offer her, studied physical training.

After spending time in New York, Steele and her husband were given some choice of where he would be stationed, and they wound up in Washington state. The Seattle area wasn't as foreign to Steele as Orlando had been, but it wasn't home either, and she describes her first two years working as a secretary and part-time personal trainer in Washington as "not great."

That soon changed—almost by accident. While investigating programs at Olympic Community College, Steele stopped a man at random to ask directions. The man, who tuned out to be the director of the school's fire science program, asked her if she would be interested.

At first, Steele rejected the idea out of hand, thinking, "I'm not a man. I'm not even big. This is something I could never do." Indeed, in a department that, Steele observes, seems to select for height—"I think the shortest man in the department is six-foot-two," she says—Steele stands out at 5 feet, 4 inches and about 150 pounds. But she was intrigued enough to mention the opportunity to her husband, whose father is in the fire service. He thought it was a good idea and encouraged Steele to try out. She surprised herself; she loved it. "I was amazed at how much I liked the physical tests," she recalls. "Getting all dirty and sweaty. It was great."

That chance encounter with Olympic's fire science department head turned Steele's life around. Since the time she was accepted into the Kitsap residents program, she says, she's been extremely happy because her colleagues at the station have become her extended family.

Kitsap County, Fire District 7

Kitsap County covers 393 square miles on the Kitsap Peninsula, across Puget Sound from Seattle. With 230,000 residents, it is the second most densely populated county in Washington and links the Seattle /

Tacoma metropolitan area with the Olympic Mountain wilderness to the west. The 16 stations of Kitsap Fire District 7 are spread over 145 square miles in the southern end of the county and serve a population of just under 70,000. As of the end of 2002, District 7 was staffed by 56 career fire fighters, 75 volunteers, and 36 volunteer residents. The district includes small city, suburban, and rural areas, as well as a considerable amount of waterfront property. Despite its large number of docks and marinas, District 7 has no dive team or fireboat, relying instead on federal fire-fighting assistance from the several military bases in the county to help with marine fires as well as with emergencies involving hazardous materials.

District 7 headquarters is Station 8 in the county seat of Port Orchard (population of roughly 7,000). Jennifer Steele lives within five minutes of the firehouse. Port Orchard is a one-hour ferry ride from Seattle, and has, in recent years, attracted commuters crowded out of the Seattle housing market. The station responds to mutual-aid calls throughout the county.

Roughly 85 percent of calls in Kitsap County are medical. Although the area is famously damp in winter, fall is wildfire season, and all Kitsap fire fighters carry pagers so they can be called in to cover the station while the bulk of a shift is off fighting brush fires.

Also each fall, there is a rush to be tested for full-time positions. Many of the station's volunteer residents never aspire to be more than part-timers, but Steele was inspired to give it try when, in 2001, a lieutenant left to become chief at another department. Everyone moved up a notch when he left, leaving an opening for one fire fighter.

The Hiring Process

Not knowing what to expect, Steele decided to be tested. She did well on her first fire fighter test, passed the physical, and shined during her oral boards. She ended up number six on the list of applicants, several of whom were her friends, and reconciled herself to not getting the job. Next came a background check and an interview with the chief. After an almost two-month hiring process, Steele was chosen out of a field that, she estimates, countywide, must have included 100 initial applicants.

"I consider myself very lucky," she says, confessing, "but I'd do this job for free."

Gender Issues

Not everyone, however, was delighted with the department's choice.

"I'd love for it not to be true," Steele says, "but being a woman in the fire service is still a factor. There's still always the suspicion that a woman was given preference over more qualified candidates because she's female."

"Being a woman in the fire service is still a factor. There's still always the suspicion that a woman was given preference over more qualified candidates because she's female."

Indeed, a ripple of gender-based resentment followed Steele's hire. One failed candidate made critical phone calls to department authorities. Relatives of a rejected candidate made comments about her qualifications that hurt Steele's feelings. Despite her enthusiasm and self-confidence, Steele herself began to wonder if she'd been the beneficiary of reverse discrimination, so she compared her qualifications to those of the other top candidates and, to her relief, determined that they were indeed equal.

"I don't know why I was chosen over other people with the same qualifications," she says. "Maybe I gave a better interview. But I have the qualifications, and in the end they could hire only one of us."

In general, Steele reports, the Pacific Northwest fire services are "female friendly." Steele does recall, though, visiting one neighboring station—one that's never had a woman on its squad—during training and discovering that the washroom door marked "female" led to a cleaning closet. But her own department, where she's currently the only paid female fire fighter (there is one full-time female paramedic, but she doesn't share Steele's shift), has been very supportive. Still, she says, it's sometimes awkward living in the station for 24 hours with five guys. She's found some of the older fire fighters to be more standoffish than the younger men are, perhaps because veterans of an earlier era can't relate to a young woman following a career path that was once exclusively male.

For the most part, Steele has become one of the boys. She's been invited along on the department's annual outing to a Seahawks game and she even trades punches in the arm with her fellow fire fighters. The gender issue, however, never completely goes away. Steele worries about being treated differently and about perceived favoritism. Fortunately, she's never encountered anything resembling sexual harassment in her department, but little things weigh on her mind.

While most of her colleagues think nothing of including her in traditionally male rituals such as mild roughhousing, a few try to intercede—"Hey, don't push her." Her chief bought her a subscription to the publication *Women in the Fire Service,* which she reads and appreciates but which is nevertheless an excuse for some ribbing.

More likely are gender problems arising with the public—with people who come to the station and insist on talking to a male fire fighter or with older male patients who are embarrassed when a woman shows up on their medical emergency call. And when it comes to a choice between lugging equipment and moving a 300-pound patient onto a gurney, Steele grabs the gear. At the same time, Steele has become quite experienced attaching cardiac monitors to female heart attack patients and, thanks to her size, being the rescuer chosen to crawl inside a wrecked car during a vehicle extrication.

Rookie Year Experience

Jennifer Steele describes her rookie year as one long—and stressful—testing process. But the stress didn't come from where she'd expected it to come. "I didn't think I'd be able to do the physical part," she says. "but I can make up for not having the brute strength with technique."

"I didn't think I'd be able to do the physical part, but I can make up for not having the brute strength with technique."

During her time as a volunteer resident, Steele's fire scene duties were largely limited to jobs like working backup on a hand line. But when she became a probationary fire fighter and hit the academy, she was held to the same physical standards as everyone else on a career track. In part because of her physical-training background, Steele was able to meet all standards, but she still worried that her size and stamina would fail her. During her rookie years, those concerns vanished. "The idea of pulling a 75-pound hose—that's half my weight—doesn't faze me," she says. "The hard part turns out to be the homework—the opposite of what I thought would be tough."

Kitsap County has high expectations for its probies, and doesn't mind making the experience stressful because, after all, stress is part of the job. Steele discovered that she had to learn specs for all sorts of equipment. Without having had a lot of mechanical background, she found herself memorizing how pumps worked and the amount of oil used in a piece of hydraulic gear. She mastered local maps in very specific detail. And she was quizzed daily on such arcana as the gas mixture for a chain-saw engine.

"I don't know if that's something I'll ever need to know as a fire fighter," she says. "But fire fighting has become more than grabbing an axe and throwing water around. More and more, fire fighters have to be educated.

"Generally when you think of fire fighters, you think of big strong burly guys, but now a lot of the training is medical. You have to read. You have to keep current. You have to keep up with medical issues."

"Fire fighting has become more than grabbing an axe and throwing water around. More and more, fire fighters have to be educated. You have to read. You have to keep current. You have to keep up with medical issues."

So amid all the requirements and obligations of the rookie experience (not to mention the stress of prepping for rigorous written and physical exams at her six-month and eleven-month marks), Steele learned to make the most of the resources at hand. She learned to ask more than one person when she wanted to learn about a piece of equipment or a procedure. "I learned that there are often two ways to do things: the way you answer for fire school and the way veteran fire fighters do things in real life."

"I also learned that different fire fighters have different ways of doing things depending on when they trained." From time to time, an old-timer would ask Steele to do something she had never heard of—a task that was once standard at the academy. At the same time, she discovered that her training included a lot of material that older veterans had never covered in fire school. By building relationships with experienced colleagues as well as with a group of young fire fighters who had been rookies just two years before, Steele was able to broaden her own education on the job—exposing herself to the theories of new and old techniques.

According to Steele, developing good relationships with colleagues at all levels was the key to a successful rookie year. "Having a good probationary year is all about the chemistry between people. Where I work, the personal chemistry is ideal. People work well together, and I'm comfortable working with them. Rookies who have a hard time are usually the ones who don't fit in. You have to have a good personality and attitude to get along. Fire fighters are constantly looking out for each other. They have to trust each other."

"Fire fighters are constantly looking out for each other. They have to trust each other."

Mistakes and Successes

In the end, the best teacher is experience, and bad experiences count as much as good ones. One night late in her rookie year, Steele was involved in the rearranging of the equipment stored on one of the department's trucks to conform with countywide standards. It was a routine job, and everyone went to bed when it was finished. At about 1:20 A.M., an alarm came in for a structure fire, but before the truck Steele was riding got to the fire scene, it was diverted back to the station for a medical emergency. As they drove back, disappointed about missing the fire, they saw their equipment strewn along the road. Apparently, someone had left the door of an equipment compartment open and the safety check mechanism that should have alerted the driver had failed.

Steele felt terrible. The equipment shift had been her responsibility, and some of the gear had been damaged when it fell out of the truck. So instead of fighting a fire, she and her lieutenant were wandering around in the dark at 2:00 A.M., retrieving broken equipment.

"That was one night I didn't sleep at all," she says. "I just lay on my bunk, thinking 'I can't believe I left the door open.' I may not have been the person who didn't close the door, but it was my responsibility. They always tell you to check your truck before you go out—do a quick "360." I don't know how many people actually do it, but I learned why it's important."

Steele's probie year was also marked by successes that won her the respect of her colleagues. Although Kitsap's volunteer rookies are required to learn to drive all the department's vehicles, when Steele signed on as a paid rookie, she'd never driven a pumper on a call. She had trained for that at the fire academy, but no amount of training matches the experience of driving a 40,000-pound, 300,000-dollar vehicle carrying 750 gallons of water to a real fire—in the dark.

Steele's first experience driving a pumper was to a garage fire, which, of course, could potentially involve unknown quantities of stored flammable liquids. The fire was called in by a passerby, so she didn't have an exact address. Her lieutenant, well aware that Steele had never driven on this sort of response, rode next to her as they followed their general directions toward the shore. He was studying the map when Steele caught sight of the fire's glow and calmly announced, "We just passed it."

When she turned around and got the truck to the fire scene, she was confronted with two big trees flanking her access. "All the time I was driving there I was nervous because I knew I'd have to pump. I kept thinking about pump-discharge procedure—going over everything I'd learned in class. Then I got there, and I had to worry about something as simple as trees in the way."

Steele got her truck through with inches to spare and began setting up. "I had trained so much," she remembers, "I went through that fire like a robot. I don't remember anything anyone said. I just did what I'd been trained to do. When my battalion chief came by, he was amazed that I appeared so calm. I felt like it had been too easy. I hadn't made any major mistakes. Everybody praised me—said I'd done a good job. It felt tremendous."

When the fire was out, the company took pictures and Steele kept a piece of burnt metal from the fire as a souvenir. "I was relieved. I'd had so much responsibility for the engine and for the people on the hand lines. My training paid off. You can't call any fire good, but if you're going to go to a fire, that was a good one."

12

Learning from Experience

Dan Volcko

Dan Volcko

PROFILE

Fire Department: Phoenix Fire Department

Location: Phoenix, Arizona

Position / Title: Fire fighter

Rookie Year: 2001–2002

Height / Weight: 5′10″ / 180 lbs.

Previous Job: Intern architect

Hobbies / Other Interests: Sports, drawing, spending time with his wife, carpentry, cooking, taking classes at the college

Education / Training: B.A. architecture, Montana State University at Bozeman; EMT-Basic; various fire service training courses, including wildland interface, extrication, defensive driving, strategy and tactics

Reason for Joining the Fire Service: "My interest in fire fighting began with a kindergarten class trip to a fire station. Since then, I can't remember a time when I didn't want to be a fire fighter."

Most Memorable Rookie Experience: "That would probably have to be putting my gear on the engine and entering my name on the roster my first shift after the academy. That's when I knew I definitely had the job."

Fire Service Ambition: "Learn, work hard, have fun, and survive the next 32 years in the fire service."

Dan Volcko had an opportunity. When he began his rookie year with the Phoenix, Arizona, Fire Department, he'd already had some fire-fighting experience, as well some invaluable exposure to the inner workings of a big city fire service. By the time he got to Phoenix, Volcko had worked with two volunteer fire organizations during the eight years he spent studying architecture in Bozeman, Montana. Later, the native of Calgary, Alberta, interned with, and then worked for, the Arizona fire department he eventually joined. Long before Volcko achieved fire fighter status, in late 2002, he had seen the fire service from several sides and knew both the harsh realities and the camaraderie that come with the job.

Volcko's volunteer stints in Montana were where, as he says, "the wide-eyed ideas of a child came to reality." Volcko, whose interest in fire fighting began with a kindergarten class trip to a fire station, can't remember a time since then when he didn't want to be a fire fighter. As a child, he built model cities just so he could drive toy fire trucks through their streets. While he was working his way through college in the States, he signed up for a residence program with a local department in Montana, where his duties included everything from janitorial work to going on calls, in exchange for the learning experience and the monetary benefit of having a place to live while in college.

The residency involved one dramatic night that tested Volcko's ability to handle the very real stresses of fire fighting. At 2:00 A.M. on a December morning, a call came in from a ski resort near town. There was light fog and some ice on the roads. As the fire truck Volcko was riding headed up the mountain, he saw car headlights swerving in the oncoming lane. The car, containing two eighteen-year-olds, spun out of control and skidded broadside across the median into the front of the fire truck, killing both the teenagers.

Volcko and the rest of the volunteers quickly checked themselves out and then ran to the kids in the car, who were already dead on impact. The accident provoked a lot of local empathy and a thorough investigation that ultimately absolved the fire department of any fault. For Volcko, the tragedy was a valuable early lesson in handling the aftermath of another individual's tragedy.

"You have to learn to confront stress and stay focused," he explains. "That's your job. [After the accident in Montana], I learned about the support network that the fire department itself provides—that taking care of each other is an every day, all-the-time culture. And I learned that you come out of a bad call the way you went in. You have to be prepared and know what to do when things change, because they will." That way, he says, "staying and working within the system provides the confidence to continue and be effective on the job and in life."

That lesson has been reinforced time and again during Volcko's work in Phoenix. "I don't know how the Phoenix police do it," he says. "They ride alone, go to traumatic events, and just get back in their separate cars after an incident. In the fire department, there's always strength in numbers. We support each other at the scene and afterwards. Being part of a good team is essential to our survival—both personally and as an organization."

"You have to learn to confront stress and stay focused. That's your job. You have to be prepared and know what to do when things change, because they will."

Phoenix Fire Department

Phoenix, Arizona, is a sprawling modern city surrounded by a dozen or so equally spread-out satellite municipalities. Unlike cramped eastern cities, designed for horses and pedestrians, most of Phoenix was developed in the past half-century and, thanks to virtually unlimited desert real estate, it's scaled to accommodate cars and trucks. To hear Phoenix natives tell it, "L.A. invented sprawl, but Phoenix perfected it."

That's a mixed blessing for the Phoenix Fire Department. Although navigating trucks and ambulances down thoroughfares that are sometimes as wide as eastern city blocks benefits response time, the city's spread-out configuration means covering greater-than-average distances between calls.

Adopting modern solutions for nontraditional problems has become a cornerstone of the Phoenix Fire Department. The department has an automatic-aid arrangement with many of the city's surrounding municipalities, which requires each of the area's departments to cover an area many times the size of its own community. The solution is an automatic vehicle locator system that can pinpoint any fire truck's current location, even if the truck is under a bridge. This allows

"We support each other at the scene and afterwards. Being part of a good team is essential to our survival—both personally and as an organization."

An automatic vehicle locator system can pinpoint any fire truck's current location, allowing centralized dispatchers to match emergency vehicles to a call, using computer technology, and dispatch whoever's closest.

centralized dispatchers to match emergency vehicles to a call, using computer technology, and dispatch whoever's closest. So a given call in Phoenix might involve a pumper from one station, an ambulance from a neighboring community, and a ladder from another.

Dan Volcko's rookie year involved typical rotations through medical and fire-fighting units, but it also exposed him to a lot of change. In fact, under Phoenix Fire Department Chief Alan V. Brunacini, change and problem solving have come to define the department. As well as learning to handle hoses, Phoenix rookies learn a workplace philosophy designed to extend their training well beyond their rookie years. Phoenix rookies learn to do everything—in and out of school—according to Chief Brunacini's training model: learn, practice, apply, and reevaluate.

Family and Background

Volcko got his introduction to Phoenix's 1,300-person fire service in the form of a paid internship. After finishing school in Montana, Volcko shopped around for a job in fire fighting. The Phoenix Fire Department is very open to civilian participation. It staffs its operations, in part, through a civilian internship program, which gives would-be fire fighters a chance to learn about the service first hand. The program attracts interns from throughout the Southwest and recruits heavily through the Oklahoma State University fire protection program. When Volcko was accepted into the program, his fiancée, Tiffany, had not yet finished school in Montana. So they were married and then spent eight months apart while Volcko worked in Arizona. The separation was difficult for the couple, but it enabled Volcko to give his full concentration to the department.

Because Volcko had taken public safety and public education courses, he was assigned to work all daylong as an intern at the Phoenix Fire Academy, running errands and helping trainers. It was very task-oriented work. If, for example, an instructor needed 50 sheets of plywood for a practical exercise, Volcko was the gofer. All the physical activity suited him fine since he'd never seen himself as a "desk-job kind of guy."

After work days, he lived in a Phoenix firehouse, where he "did all the things my mother taught me to do to be a good guest." He helped out with chores, did dishes, and, as a civilian, actually went out on calls to observe and help out when possible. On top of his all-day duties at the academy, he had a grueling schedule. But the experience convinced Volcko that he wanted to work in Phoenix and, specifically, that he wanted to work under Chief Brunacini, whose openness to change and determination to make the department's biggest investment be in its fire fighters impressed him. "There were posters around the Phoenix stations that quoted Chief Brunacini," Volcko recalls, "that said something to the effect that 'We used to put more money into our trucks than into our people.' "

In practical terms, Volcko's internship months were a tremendous asset. It showed Volcko the big picture of the Phoenix Fire Department. It gave him an overview of how the department worked, from equipping trucks to making sure the stations had enough paper towels.

Testing and Hiring Process

After Dan's internship, Dan and Tiffany went on their honeymoon and relocated together in a condo in downtown Phoenix. For the next year, Volcko worked at the academy where he'd interned—this time as a full-time civil servant, with benefits. It was during that year that he applied to be a recruit at the academy. The process took about nine months.

A fellow Phoenix rookie once described getting a job as a fire fighter as going on a scavenger hunt when you don't know what's on the list. That may be an exaggeration, but Volcko certainly faced a lot of competition for a slot at the academy. The Phoenix Fire Department accepts rookie applications for one week each January. Six to seven weeks after that, everyone who's applied takes an entrance exam at the same time.

The morning of the test, Volcko showed up early to find a long line snaking around the civic plaza; roughly 2500 people were there looking for about a dozen jobs. Many had been there since 7:30A.M. because if you're late for the test, you don't get in.

All those applicants took the test in the same room. "I was in a giant room surrounded by a sea of people who wanted 'my' job,"

"[When taking the Phoenix Fire Department entrance exam] I was in a giant room surrounded by a sea of people who wanted 'my' job."

Volcko recalls. Test scores aren't released for months, at which point the lowest-scoring third fail and the top third are each interviewed by a city employee, a fire captain, and a fire fighter. (The middle third are put on hold, to be interviewed later when the department selects the academy class that will begin study the next January.) Six to eight weeks after the first interview round, applicants are interviewed again before being accepted or rejected. Volcko scored well on the test and was accepted into the fall 2001 class.

Phoenix Fire Department's Alternative Resource Program

Volcko's record in Montana and his experience working with the Phoenix department were certainly pluses, but, in fact, most academy students, he reports, have done volunteer residencies in other departments or even taken Fire Fighter I and II courses on their own at a community college. Getting hired really means getting involved, learning the organization, and getting to know the job. This is true of any profession.

Still others participated in the Phoenix Fire Department's Alternative Resource (AR) program, which fields volunteers, mostly college students and EMTs, to provide on-the-scene counseling to the victims of fire or medical emergencies. These volunteers are just one aspect of the Phoenix Fire Department's dedication to what can best be described as customer service. After a disaster, people often have no idea what to do. The Alternative Resource volunteers give advice about everything from follow-up medical treatment and funeral arrangements to disposing of a wrecked car or explaining why fire fighters are chainsawing holes in someone's roof.

The Alternative Resource volunteers give advice about everything from follow-up medical treatment and funeral arrangements to disposing of a wrecked car or explaining why fire fighters are chainsawing holes in someone's roof.

Although Volcko himself didn't participate as an AR volunteer, the program, and similar initiatives that Chief Brunacini refers to as "value-added service," are a large part of what attracted Volcko to the Phoenix Fire Department. Brunacini encourages his staff to help people deal with the practical details that accompany a tragedy. In one case, Volcko reports, a fire fighter finished giving a haircut to a man whose barber had had a heart attack while the customer was in the chair.

"Helping victims and doing a little more has become part of the department culture," Volcko explains. "It's almost as if each [response] team tries to outdo the other guys with their own inventiveness. Simply put, life and our job are a lot easier when we are nice."

"Helping victims and doing a little more has become part of the department culture. It's almost as if each [response] team tries to outdo the other guys with their own inventiveness."

Rookie Year Experience

During one of Volcko's rookie ambulance shifts, for example, he was called to the Phoenix airport to intercept a woman who had been headed for a rehab center in Palm Springs but who had drunk herself sick on the plane. After the ambulance team took the woman to a hospital, they turned their attention to her two female companions who found themselves unexpectedly stranded in a strange city. The team gave the women a cell-phone number to call when they'd settled their friend at the hospital, then drove the women back to the airport to retrieve their luggage, and escorted them to a hotel.

"It all becomes part of the job," says Volcko. "You don't leave people high and dry. You try to understand the complexity of each person's emergency."

Community relations was an undercurrent throughout Volcko's rookie experience, and for good reason. Phoenix can be a tough city for police. There are parts of the city with large poor populations and high crime rates. Language barriers, cultural differences, and economic disparity, coupled with the high availability of firearms, make police work physically dangerous. Relations between cops and the communities they protect are sometimes strained, so the fire service has gone out of its way to differentiate fire fighters from law enforcement personnel. Phoenix fire fighters wear T-shirts, not uniforms with badges, and work hard to maintain a positive relationship in the city's neighborhoods.

Volcko had the opportunity to witness the importance of community relations during his first rookie rotation, when his team was called

"When a victim codes, there's a lot of frantic-looking EMT activity that looks threatening. That's why we devote resources and people to deal with the victim's family. Our job is to help, but it's also to teach and interact."

to a predominately Spanish-speaking community where a man had been stabbed. "Everybody wanted us to help," Volcko recalls, "but culturally, they didn't understand the EMT's procedures. The bystanders were upset and excited. Their involvement could have interfered with our ability to help the victim. For instance, when a victim codes, there's a lot of frantic-looking EMT activity that looks threatening. That's why we devote resources and people to deal with the victim's family. Our job is to help, but it's also to teach and interact."

Volcko's rookie year began with twelve weeks of training at the Phoenix Regional Training Academy where he interned and worked. Academy training involved one week covering medical response, which makes up the majority of any fire fighter's calls, and eleven weeks on, as Volcko calls them, "the situations that can hurt you."

After that, a Phoenix rookie spend nine months rotating through ladder, ALS, and BLS engine companies around the city. The Phoenix Fire Department is by nature fluid. Although there are core groups that are permanently assigned to individual stations, a large number of Phoenix fire fighters are "rovers," whose typical day begins with a 6:00 A.M.-phone call to headquarters to find out where they're assigned for their shift. So rookies have to get used to moving around the department and adapting to a variety of work situations. Volcko spent two of his three rookie rotations assigned to basic lifesaving BLS duties on EMT trucks and a third with a ladder company.

The Phoenix Fire Department responds to approximately 135,000 calls a year. Including mutual-aid calls with Phoenix's surrounding communities, that number jumps to between 200,000 and 300,000. And the majority of those calls—roughly 80 percent—are medical.

To meet that demand and to give new fire fighters as much experience on medical calls as possible, each new fire fighter is required to do 200 ambulance shifts before they get back on a fire truck. With an average post-rookie workload of 120 shifts a year, the ambulance requirement can take up to two years to fulfill.

Volcko's first rookie assignment was to an EMT group. He was the fourth man in the company and his colleagues expected him to function up to speed. Volcko describes his very first response as a minor car accident involving no serious injuries. Still, it offered him a further introduction to the department's emphasis on customer service. "I learned right away that a lot of the department's job is like social work," says Volcko. "The patient involved—who ended up leaving the scene on his own—thought it was a major emergency, and to *him* it was. He wasn't physically hurt, so then we dealt with his anxiety and stress of being in a car accident. People want understanding. The department takes those needs seriously."

"I learned right away that a lot of the department's job is like social work. The patient [of a minor accident] thought it was a major emergency, and to him *it was. He wasn't physically hurt, so then we dealt with his anxiety and stress of being in a car accident. The department takes those needs seriously."*

For his second rookie rotation, Volcko worked as the fifth man on a downtown ladder company. A fifth-man rookie on a ladder truck is treated as essential personnel. Volcko's captain, a seasoned, thirty-six-year veteran, pushed Volcko to do things and get hands-on experience. For his third round, Volcko went back to BLS work on another downtown engine company.

The most dramatic call of Volcko's rookie year involved rescue duty but no rescue. Volcko was on the second of three trucks to arrive at a fire in an abandoned single-family home and was ordered to form a rescue crew to back up fire fighters who were going to enter the building to look for squatters. The fire fighters were inside when their hose broke. Fortunately, they were able to get themselves out before Volcko's rescue team, armed with tools and a thermal imaging camera, got to the building, Volcko's group administered first aid to the injured, without having to enter the burning building. But the lesson Volcko took away from that emergency was probably more important than if he'd had to help his fellow fire fighters out of the fire.

"I was impressed by everyone's discipline and dedication to the command system. When the line broke, everybody stayed in their positions. They didn't all rush in to the rescue like moths to a candle. They did their jobs."

During the course of his rookie year, Volcko came to the realization that fire service training involves more than applying skill learned at the academy. An important part of good training, he believes, is being encouraged to learn from each experience—large or small, good or bad. Under Chief Brunacini, the process of constant reevaluation has become second nature. On the large scale, the department sets the example by revising theoretical procedures in reaction to real-life events.

"I was impressed by everyone's discipline and dedication to the command system. When the line broke, everybody stayed in their positions. They didn't all rush in to the rescue. . . . They did their jobs."

When a Phoenix fire fighter was killed recently fighting a supermarket fire, for example, the department investigated the response for months and issued a lengthy report on the circumstances that resulted in the fire fighter's death. Then, even more importantly, the department created a new model for procedure based on the facts of that incident. They produced training videos that demonstrated the ideal approach and reactions in similar situations. Then the learned activities were repeatedly trained in a hands-on, in-context environment. Entire procedures were improved as a result.

On a smaller scale, vehicles dispatched to medical calls now park with a plan—usually an ambulance first, with a fire truck behind—so as to protect medics as they move gurneys and victims on and off the ambulance. This procedure resulted from one tragic death in 1994, when an ambulance that had arrived at an emergency scene second and was parked behind the fire truck was hit by a drunk driver as the EMTs were loading the patient into the ambulance.

The learn-from-reality approach doesn't end with official department-wide policy changes. Part of Volcko's rookie indoctrination involved growing into a level of responsibility that the Phoenix Fire Department expects of its fire fighters. Chief Brunacini calls it "fire fighter empowerment." Faced with unusual circumstances or needs, fire fighters are told, in effect, "Don't ask; just do it." If something is not illegal or harmful and makes sense at the time, Phoenix fire fighters are expected to act on their own initiative.

Fresh off his rookie, or "booter," year, Volcko was amazed at the feeling of relief. "When you're on probation," he says, "you're expected to listen and learn. You're being evaluated and under constant scrutiny all the time. Even though you might not realize it at the time, the stress level is very high. Applying newly learned skills and just fitting in day to day into a unique culture add a lot of anxiety."

"When you're on probation, you're expected to listen and learn. You're being evaluated and under constant scrutiny all the time. The stress level is very high."

It's exhausting. When I was a rookie, I was always tired. I didn't understand how the veterans handled having second jobs—

often physically challenging jobs in construction. Then when I went off probation, I realized that it was the stress that wore me down. Now I can relax, do my job, and be myself."

"You need a balanced lifestyle. Fire fighting is the best job I have ever had, but it's also the other things you do, the company you keep, and the way you treat people that make you who you are."

A fire fighter's life involves a lot of time management. Working twenty-four-hour shifts separated by forty-eight hours off challenges fire fighters to make good use of their time away from the fire service. Volcko plays sports, draws, spends time with his wife, does carpentry, and takes classes. He finds that diversity and doing things you enjoy are important parts of living a full life.

"Fire fighting is not unlike other professions where people have become involved through their passions. Sometimes stepping back and taking time to enjoy life helps prolong a healthy career. You can get in the trap of having your entire self-esteem come from being a fire fighter. You need a balanced lifestyle. Fire fighting is the best job I have ever had, but it's also the other things you do, the company you keep, and the way you treat people that make you who you are."

APPENDIX A

Career Opportunities for Fire Fighters, EMTs, and Paramedics

Fire can be one of the most dangerous and costly of all destructive forces. In response to the hazard that fire presents to both life and property, the scientific study of fire protection and the methods of fire prevention have grown since the early nineteenth century. From this growth and from recent technological advances in building design and the increasing use of hazardous materials, the fire protection profession has become highly specialized and challenging.

A student of the fire sciences is presented with a variety of career opportunities, including the professions of fire fighter, emergency medical technician (EMT) or paramedic, fire prevention educator, fire investigator or fire inspector, as well as positions in many fire-related organizations and fields such as fire insurance businesses, designers and manufacturers of fire equipment, industrial engineering, and other opportunities. For the purposes of this book, however, we will limit our discussion to career opportunities for fire fighters and—because many fire departments find that medical calls now exceed fire calls—opportunities for EMTs and paramedics.

Fire-Fighting Occupations*

Significant Points

- Fire fighting involves hazardous conditions and long, irregular hours.

**Source:* Bureau of Labor Statistics, U.S. Department of Labor, *Occupational Outlook Handbook, 2002–03 Edition,* Fire-Fighting Occupations, on the Internet at *http://www.bls.gov/oco/ocos158.htm.*

- Keen competition for jobs is expected; many people are attracted to the occupation because it provides considerable job security and the opportunity to perform an essential public service.

Nature of the Work

Every year, fires and other emergencies take thousands of lives and destroy property worth billions of dollars. Fire fighters help protect the public against these dangers by rapidly responding to a variety of emergencies. They are frequently the first emergency personnel at the scene of a traffic accident or medical emergency and may be called upon to put out a fire, treat injuries, or perform other vital functions.

During duty hours, fire fighters must be prepared to respond immediately to a fire or any other emergency that arises. Since fighting fires is dangerous and complex, it requires organization and teamwork. At every emergency scene, fire fighters perform specific duties assigned by a superior officer. At actual fires, they connect hose lines to hydrants, operate a pump to send water to high pressure hoses, and position ladders to enable them to deliver water to the fire. They also rescue victims and provide emergency medical attention as needed, ventilate smoke-filled areas, and attempt to salvage the contents of buildings. Their duties may change several times while the company is in action. Sometimes they remain at the site of a disaster for days at a time, rescuing trapped survivors and assisting with medical treatment.

Fire fighters have assumed a range of responsibilities, including emergency medical services. In fact, most calls to which fire fighters respond involve medical emergencies, and about half of all fire departments provide ambulance service for victims. Fire fighters receive training in emergency medical procedures, and many fire departments require them to be certified as *emergency medical technicians* (see the discussion on EMTs and paramedics on p. 137.)

Fire fighters work in a variety of settings, including urban and suburban areas, airports, chemical plants, other industrial sites, and rural areas such as grasslands and forests. In addition, some fire fighters work in hazardous materials units that are trained for the control, prevention, and cleanup of oil spills and other hazardous materials incidents. Workers in urban and suburban areas, airports, and industrial sites typically use conventional fire-fighting equipment and tactics, while forest fires and major hazardous materials spills call for different methods.

In national forests and parks, *forest fire inspectors* and *prevention specialists* spot fires from watchtowers and report their findings to headquarters by telephone or radio. Forest rangers patrol to ensure travelers and campers comply with fire regulations. When fires break out, crews of fire fighters are brought in to suppress the blaze, using heavy equipment, hand tools, and water hoses. Forest fire fighting, like urban fire fighting, can be rigorous work. One of the most effective means of battling the blaze is by creating fire lines by cutting down trees and digging out grass and all other combustible vegetation, creating bare land in the path of the fire that deprives it of fuel. Elite fire fighters, called *smokejumpers,* parachute from airplanes to reach otherwise inaccessible areas. This can be extremely hazardous because the crews have no way to escape if the wind shifts and causes the fire to burn toward them.

Between alarms, fire fighters clean and maintain equipment, conduct practice drills and fire inspections, and participate in physical fitness activities. They also prepare written reports on fire incidents and review fire science literature to keep abreast of technological developments and changing administrative practices and policies.

Most fire departments have a fire prevention division, usually headed by a *fire marshal* and staffed by *fire inspectors* and *fire and life safety educators*. Workers in this division conduct inspections of structures to prevent fires and ensure fire code compliance. These fire fighters also work with developers and planners to check and approve plans for new buildings. Fire prevention personnel often speak on fire and life safety in schools and before public assemblies and civic organizations.

Some fire fighters become *fire investigators* who determine the origin and causes of fires. They collect evidence, interview witnesses, and prepare reports on fires in cases where the cause may be arson or criminal negligence. They often are called upon to testify in court.

Working Conditions

Fire fighters spend much of their time at fire stations, which usually have features common to a residential facility such as a dormitory. When an alarm sounds, fire fighters respond rapidly, regardless of the weather or hour. Fire fighting involves risk of death or injury from sudden cave-ins of floors, toppling walls, traffic accidents when responding to calls, and exposure to flames and smoke. Fire fighters may also come in contact with poisonous, flammable, or explosive gases and chemicals, as well as radioactive or other hazardous materials that may have immediate or long-term effects on their health. For these reasons, they must wear protective gear that can be very heavy and hot.

Work hours of fire fighters are longer and vary more widely than hours of most other workers. Many fire fighters work more than 50 hours a week, and sometimes they may work even longer. In some agencies, they are on-duty for 24 hours, then off-duty for 48 hours, and receive an extra day off at intervals. In other departments, they work a day shift of 10 hours for 3 or 4 days; a night shift of 14 hours for 3 or 4 nights; have 3 or 4 days off, and then repeat the cycle. In addition, fire fighters often work extra hours at fires and other emergencies and are regularly assigned to work on holidays. Fire lieutenants and fire captains often work the same hours as the fire fighters they supervise. Duty hours include time when fire fighters study, train, and perform fire prevention duties.

Employment

Employment figures in this statement from the U.S. Department of Labor's *Occupational Outlook Handbook* include only those for paid career fire fighters—they do not cover volunteer fire fighters, who perform the same duties and may comprise the majority of fire fighters in a residential area. Paid career fire fighters held about 258,000 jobs in the year 2000. First-line supervisors / managers of fire-fighting and prevention workers held about 62,000 jobs; and fire inspectors held about 13,000. More than 9 out of 10 employees worked in municipal or county fire departments. Some large cities have thousands of career fire fighters, while many small towns have only a few. Most of the remainder worked in fire departments on federal and state installations, including airports. Private fire-fighting companies employ a small number of fire fighters and usually operate on a subscription basis.

In response to the expanding role of fire fighters, some municipalities have combined fire prevention, public fire education, safety, and emergency medical services into a single organization commonly referred to as a public safety organization. Some local and regional fire departments are being consolidated into countywide establishments in order to reduce administrative staffs and cut costs, and to establish consistent training standards and work procedures.

Training, Other Qualifications, and Advancement

Applicants for municipal fire-fighting jobs generally must pass a written exam; tests of strength, physical stamina, coordination, and agility; and a medical examination that includes drug screening. Workers may be monitored on a random basis for drug use after accepting employment. Examinations are generally open to persons who are at least 18 years of age and have a high school education or the equivalent. Those who receive the highest scores in all phases of testing have the best chances for appointment. The completion of community college courses in fire science may improve an applicant's chances for appointment. In recent years, an increasing proportion of entrants to this occupation has had some postsecondary education.

As a rule, entry-level workers in large fire departments are trained for several weeks at the department's training center or academy. Through classroom instruction and practical training, the recruits study fire-fighting techniques, fire prevention, hazardous materials control, local building codes, and emergency medical procedures, including first aid and cardiopulmonary resuscitation. They also learn how to use axes, chain saws, fire extinguishers, ladders, and other fire-fighting and rescue equipment. After successfully completing this training, they are assigned to a fire company, where they undergo a period of probation.

A number of fire departments have accredited apprenticeship programs lasting up to 5 years. These programs combine formal, technical instruction with on-the-job training under the supervision of experienced fire fighters. Technical instruction covers subjects such as fire-fighting techniques and equipment, chemical hazards associated with various combustible building materials, emergency medical procedures, and fire prevention and safety. Fire departments frequently conduct training programs, and some fire fighters attend advanced training sessions sponsored by the U.S. National Fire Academy. These training sessions cover topics including executive development, anti-arson techniques, disaster preparedness, hazardous materials control, and public fire safety and education. Some states also have extensive fire-fighter training and certification programs. In addition, a number of colleges and universities offer courses leading to 2- or 4-year degrees in fire engineering or fire science. Many fire departments offer fire fighters incentives such as tuition reimbursement or higher pay for completing advanced training.

Among the personal qualities fire fighters need are mental alertness, self-discipline, courage, mechanical aptitude, endurance, strength, and a sense of public service. Initiative and good judgment are also extremely important because fire fighters make quick decisions in emergencies. Since members of a crew live and work closely together under conditions of stress and danger for extended periods, they must be dependable and able to get along well with others. Leadership qualities are necessary for officers, who must establish and maintain discipline and efficiency, as well as direct the activities of fire fighters in their companies.

Most experienced fire fighters continue studying to improve their job performance and prepare for promotion examinations. To progress to higher-level positions, they acquire expertise in advanced fire-fighting equipment and techniques, building construction, emergency medical technology, writing, public speaking, management and budgeting procedures, and public relations.

Opportunities for promotion depend upon written examination results, job performance, interviews, and seniority. Increasingly, fire departments use assessment centers that simulate a variety of actual job performance tasks to screen the best candidates for promotion. The line of promotion usually is to engineer, lieutenant, captain, battalion chief, assistant chief, deputy chief, and finally to chief. Many fire departments now require a bachelor's degree, preferably in fire science, public administration, or a related field, for promotion to positions higher than battalion chief. A master's degree is required for executive fire officer certification from the National Fire Academy and for state chief officer certification.

Job Outlook

Prospective fire fighters are expected to face keen competition for available job openings. Many people are attracted to fire fighting because it is challenging and provides the opportunity to perform an essential public service, a high school education is usually sufficient for entry, and a pension is guaranteed upon retirement after 20 years. Consequently, the number of qualified applicants in most areas exceeds the number of job openings, even though the written examination and physical requirements eliminate many applicants. This situation is expected to persist in coming years.

Employment of fire fighters is expected increase more slowly than the average for all occupations through 2010 as fire departments continue to compete with other public safety providers for funding. Most job growth will occur as volunteer fire-fighting positions are converted to paid positions. In addition to job growth, openings are expected to result from the need to replace fire fighters who retire, stop working for other reasons, or transfer to other occupations.

Layoffs of fire fighters are uncommon. Fire protection is an essential service, and citizens are likely to exert considerable pressure on local officials to expand or at least preserve the level of fire protection. Even when budget cuts do occur, local fire departments usually cut expenses by postponing equipment purchases or not hiring new fire fighters, rather than by laying off staff.

Earnings

Median hourly earnings of fire fighters were $16.43 in 2000. The middle 50 percent earned between $11.82 and $21.75 per hour. The lowest 10 percent earned less than $8.03, and the highest 10 percent earned more than $26.58 per hour. Median hourly earnings were $16.71 in local government and $15.00 in federal government.

Median annual earnings of first-line supervisors / managers of fire fighting and prevention workers were $51,990 in 2000. The middle 50 percent earned between $40,920 and $64,760. The lowest 10 percent earned less than $31,820, and the highest 10 percent earned more than $77,700. First-line supervisors / managers of fire fighting and prevention workers employed in local government earned about $52,390 a year in 2000.

Median annual earnings of fire inspectors and investigators were $41,630 in 2000. The middle 50 percent earned between $31,630 and $53,130 a year. The lowest 10 percent earned less than $24,790, and the highest 10 percent earned more than $65,030. Fire inspectors and investigators employed in local government earned about $44,030 a year in 2000.

Median annual earnings of forest fire inspectors and prevention specialists were $32,140 in 2000. The middle 50 percent earned between $22,930 and $41,150 a year. The lowest 10 percent earned less than $17,060, and the highest 10 percent earned more than $50,680.

The International City-County Management Association's annual Police and Fire Personnel, Salaries, and Expenditures Survey revealed that 89 percent of the municipalities surveyed provided fire protection services in 2000. The following 2000 salaries pertain to sworn full-time positions.

	Minimum annual average base salary	*Maximum annual average base salary*
Fire chief	$58,156	$74,749
Deputy chief	52,174	65,112
Battalion chief	50,164	62,309
Assistant fire chief	48,391	60,179
Fire captain	41,816	50,848
Fire lieutenant	38,875	46,327
Fire prevention/code inspector	37,142	46,798
Engineer	35,090	44,310
Fire fighter	29,316	39,477

Fire fighters who average more than a certain number of hours a week are required to be paid overtime. The hours threshold is determined by the department during the fire fighter's work period, which ranges from 7 to 28 days. Fire fighters often earn overtime for working extra shifts to maintain minimum staffing levels or for special emergencies.

Fire fighters receive benefits that usually include medical and liability insurance, vacation and sick leave, and some paid holidays. Almost all fire departments provide protective clothing (helmets, boots, and coats) and breathing apparatus, and many also provide dress uniforms. Fire fighters are generally covered by pension plans, often providing retirement at half pay after 25 years of service or if disabled in the line of duty.

*Source of Additional Information**

Information about a career as a fire fighter may be obtained from local fire departments and from:

- **International Association of Fire Fighters**
  1750 New York Ave., NW
  Washington, DC 20006
  Internet: *http://www.iaff.org*

- **U.S. Fire Administration**
  16825 South Seton Ave.
  Emmitsburg, MD 21727
  Internet: *http://www.usfa.fema.gov*

Information about fire fighter professional qualifications and degree programs in fire science or fire prevention may be obtained from:

- **National Fire Academy, Degrees at a Distance Program**
  16825 South Seton Ave.
  Emmitsburg, MD 21727
  Internet: *http://www.usfa.fema.gov/dhtml/fire-service/tr_ddp_enrl.cfm*

An industry employing fire-fighting occupations that appears in the 2002-03 *Career Guide to Industries*: State and local governments, except education and health.

**Disclaimer:* Links to non-BLS Internet sites are provided for your convenience and do not constitute an endorsement.

Emergency Medical Technicians and Paramedics*

Significant Points

- Job stress is common because of irregular hours and treating patients in life-or-death situations.
- Formal training and certification are required but state requirements vary.
- Employment is projected to grow faster than average as paid emergency medical technician positions replace unpaid volunteers.

Nature of the Work

People's lives often depend on the quick reaction and competent care of emergency medical technicians (EMTs) and paramedics, who are EMTs with additional advanced training to perform more difficult prehospital medical procedures. Incidents as varied as automobile accidents, heart attacks, drownings, childbirth, and gunshot wounds all require immediate medical attention. EMTs and paramedics provide this vital attention as they care for and transport the sick or injured to a medical facility.

Depending on the nature of the emergency, EMTs and paramedics typically are dispatched to the scene by a 911 operator and often work with police and fire department personnel. Once they arrive, they determine the nature and extent of the patient's condition while trying to ascertain whether the patient has preexisting medical problems. Following strict rules and guidelines, they give appropriate emergency care and, when necessary, transport the patient. Some paramedics are trained to treat patients with minor injuries on the scene of an accident or at their home without transporting them to a medical facility. Emergency treatments for more complicated problems are carried out under the direction of medical doctors by radio preceding or during transport.

EMTs and paramedics may use special equipment such as backboards to immobilize patients before placing them on stretchers and securing them in the ambulance for transport to a medical facility. Usually, one EMT or paramedic drives while the other monitors the

---

**Source:* Bureau of Labor Statistics, U.S. Department of Labor, *Occupational Outlook Handbook, 2002–03 Edition*, Emergency Medical Technicians and Paramedics, on the Internet at *http://www.bls.gov/oco/ocos101.htm*.

patient's vital signs and gives additional care as needed. Some EMTs work as part of the flight crew of helicopters that transport critically ill or injured patients to hospital trauma centers.

At the medical facility, EMTs and paramedics help transfer patients to the emergency department, report their observations and actions to staff, and may provide additional emergency treatment. After each run, EMTs and paramedics replace used supplies and check equipment. If a transported patient had a contagious disease, EMTs and paramedics decontaminate the interior of the ambulance and report cases to the proper authorities.

Beyond these general duties, the specific responsibilities of EMTs and paramedics depend on their levels of qualification and training. To determine this, the National Registry of Emergency Medical Technicians (NREMT) registers emergency medical-service (EMS) providers at four levels: First Responder, EMT-Basic, EMT-Intermediate, and EMT-Paramedic. Some states, however, do their own certification and use numeric ratings from 1 to 4 to distinguish levels of proficiency.

The lowest level—First Responders—are trained to provide basic emergency medical care because they tend to be the first persons to arrive at the scene of an incident. Many fire fighters, police officers, and other emergency workers have this level of training. The EMT-Basic, also known as EMT-1, represents the first component of the emergency medical technician system. An EMT-1 is trained to care for patients on accident scenes and on transport by ambulance to the hospital under medical direction. The EMT-1 has the emergency skills to assess a patient's condition and manage respiratory, cardiac, and trauma emergencies.

The EMT-Intermediate (EMT-2 and EMT-3) has more advanced training that allows administration of intravenous fluids, use of manual defibrillators to give lifesaving shocks to a stopped heart, and use of advanced airway techniques and equipment to assist patients experiencing respiratory emergencies. EMT-Paramedics (EMT-4) provide the most extensive pre-hospital care. In addition to the procedures already described, paramedics may administer drugs orally and intravenously, interpret electrocardiograms (EKGs), perform endotracheal intubations, and use monitors and other complex equipment to assess injuries.

Working Conditions

EMTs and paramedics work both indoors and outdoors, in all types of weather. They are required to do considerable kneeling, bending, and heavy lifting. These workers risk noise-induced hearing loss from sirens and back injuries from lifting patients. In addition, EMTs and paramedics may be exposed to diseases such as Hepatitis-B and AIDS, as well as violence from drug overdose victims or mentally unstable patients. The work is not only physically strenuous, but also stressful, involving life-or-death situations and suffering patients. Nonetheless, many people find the work exciting and challenging and enjoy the opportunity to help others.

EMTs and paramedics employed by fire departments work about 50 hours a week. Those employed by hospitals frequently work between 45 and 60 hours a week, and those in private ambulance services, between 45 and 50 hours. Some of these workers, especially those in police and fire departments, are on call for extended periods. Emergency services function 24 hours a day and EMTs and paramedics have irregular working hours that add to job stress.

Employment

EMTs and paramedics held about 172,000 jobs in 2000. Most career EMTs and paramedics work in metropolitan areas. There are many more volunteer EMTs and paramedics, especially in smaller cities, towns, and rural areas. They volunteer for fire departments, emergency medical services (EMS), or hospitals and may respond to only a few calls for service per month, or may answer the majority of calls, especially in smaller communities. EMTs and paramedics work closely with fire fighters, who often are certified as EMTs as well and act as first responders.

Full- and part-time paid EMTs and paramedics were employed in a number of industries. About 4 out of 10 people worked in local and suburban transportation as employees of private ambulance services. About 3 out of 10 employees worked in local government for fire departments, public ambulance services, and EMS. Another 2 out 10 were found in hospitals where they worked full time within the medical facility or responded to calls in ambulances or helicopters to transport critically ill or injured patients. The remainder worked in various industries providing emergency services.

Training, Other Qualifications, and Advancement

Formal training and certification is needed to become an EMT or a paramedic. All 50 states possess a certification procedure. In 38 states and the District of Columbia, registration with the National Registry of Emergency Medical Technicians (NREMT) is required at some or all levels of certification. Other states administer their own certification examination or provide the option of taking the NRMET examination. To maintain certification, EMTs and paramedics must re-register, usually every 2 years. In order to re-register, an individual must be working as an EMT or paramedic and meet a continuing education requirement.

Training is offered at progressive levels: EMT-Basic, also known as EMT-1; EMT-Intermediate, or EMT-2 and EMT-3; and EMT-Paramedic, or EMT-4. The EMT-Basic represents the first level of skills required to work in the emergency medical system. Coursework typically emphasizes emergency skills such as managing respiratory, trauma, and cardiac emergencies and patient assessment. Formal courses are often combined with time in an emergency room or ambulance. The program also provides instruction and practice in dealing with bleeding, fractures, airway obstruction, cardiac arrest, and emergency childbirth. Students learn to use and maintain common emergency equipment, such as backboards, suction devices, splints, oxygen delivery systems, and stretchers. Graduates of approved EMT basic training programs who pass a written and practical examination administered by the state certifying agency or the NREMT earn the title of Registered EMT-Basic. The course also is a prerequisite for EMT-Intermediate and EMT-Paramedic training.

EMT-Intermediate training requirements vary from state to state. Applicants can use the option to receive training in EMT-Shock Trauma, where the caregiver learns to start intravenous fluids and give certain medications, or in EMT-Cardiac, which includes learning heart rhythms and administering advanced medications. Training commonly includes 35 to 55 hours of additional instruction beyond EMT-Basic coursework and covers patient assessment, as well as the use of advanced airway devices and intravenous fluids. Prerequisites for taking the EMT-Intermediate examination include registration as an EMT-Basic, required classroom work, and a specified amount of clinical experience.

The most advanced level of training for this occupation is EMT-Paramedic. At this level, the caregiver receives additional training in body function and more advanced skills. The Paramedic Technology program usually lasts up to two years and results in an associate degree in applied science. Such education prepares the graduate to take the NREMT examination and become certified as an EMT-Paramedic. Extensive related coursework and clinical and field experience is required. Due to the longer training requirement, almost all EMT-Paramedics are in paid positions. Refresher courses and continuing education are available for EMTs and paramedics at all levels.

EMTs and paramedics should be emotionally stable, have good dexterity, agility, and physical coordination, and be able to lift and carry heavy loads. They also need good eyesight (corrective lenses may be used) with accurate color vision.

Advancement beyond the EMT-Paramedic level usually means leaving fieldwork. An EMT-Paramedic can become a supervisor, operations manager, administrative director, or executive director of emergency services. Some EMTs and paramedics become instructors, dispatchers, or physician assistants; others move into sales or marketing of emergency medical equipment. A number of people become EMTs and paramedics to assess their interest in healthcare and then decide to return to school and become registered nurses, physicians, or other health workers.

Job Outlook

Employment of emergency medical technicians and paramedics is expected to grow faster than the average for all occupations through 2010. Population growth and urbanization will increase the demand for full-time paid EMTs and paramedics rather than for volunteers. In addition, a large segment of the population—the aging baby boomers—will further spur demand for EMT services, as they become more likely to have medical emergencies. There will still be demand for part-time, volunteer EMTs and paramedics in rural areas and smaller metropolitan areas. In addition to job growth, openings will occur from replacement needs; some workers leave because of stressful working conditions, limited advancement potential, and the modest pay and benefits in the private sector.

Most opportunities for EMTs and paramedics are expected to arise in hospitals and private ambulance services. Competition will be greater for jobs in local government, including fire, police, and independent third service rescue squad departments, where salaries and benefits tend to be slightly better. Opportunities will be best for those who have advanced certifications, such as EMT-Intermediate and EMT-Paramedic, as clients and patients demand higher levels of care before arriving at the hospital.

Earnings

Earnings of EMTs and paramedics depend on the employment setting and geographic location as well as the individual's training and experience. Median annual earnings of EMTs and paramedics were $22,460 in 2000. The middle 50 percent earned between $17,930 and

$29,270. The lowest 10 percent earned less than $14,660, and the highest 10 percent earned more than $37,760. Median annual earnings in the industries employing the largest numbers of EMTs and paramedics in 2000 were:

Local government	$24,800
Hospitals	23,590
Local and suburban transportation	20,950

Those in emergency medical services who are part of fire or police departments receive the same benefits as fire fighters or police officers. For example, many are covered by pension plans that provide retirement at half pay after 20 or 25 years of service or if disabled in the line of duty.

*Sources of Additional Information**

General information about emergency medical technicians and paramedics is available from:

- **National Association of Emergency Medical Technicians**
  408 Monroe St.
  Clinton, MS 39056
  Internet: *http://www.naemt.org*
- **National Registry of Emergency Medical Technicians**
  P.O. Box 29233
  Columbus, OH 43229
  Internet: *http://www.nremt.org*
- **National Highway Transportation Safety Administration**
  EMS Division
  400 7th St. SW., NTS-14
  Washington, DC
  Internet: *http://www.nhtsa.dot.gov/people/injury/ems*

Selected industries employing emergency medical technicians and paramedics that appear in the 2002–03 *Career Guide to Industries*

**Disclaimer:* Links to non-BLS Internet sites are provided for your convenience and do not constitute an endorsement.

APPENDIX

B

NFPA 1001, *Standard for Fire Fighter Professional Qualifications*, 2002 Edition

NOTICE: An asterisk (*) following the number or letter designating a paragraph indicates that explanatory material on the paragraph can be found in Annex A.

A reference in brackets [ ] following a section or paragraph indicates material that has been extracted from another NFPA document. As an aid to the user, Annex D lists the complete title and edition of the source documents for both mandatory and nonmandatory extracts. Editorial changes to extracted material consist of revising references to an appropriate division in this document or the inclusion of the document number with the division number when the reference is to the original document. Requests for interpretations or revisions of extracted text shall be sent to the appropriate technical committee.

Information on referenced publications can be found in Chapter 2 and Annex D.

Chapter 1 Administration

1.1 Scope. This standard identifies the minimum job performance requirements for career and volunteer fire fighters whose duties are primarily structural in nature.

1.2 Purpose. The purpose of this standard is to specify the minimum job performance requirements for fire fighters. It is not the intent of the standard to restrict any jurisdiction from exceeding these requirements.

1.3 General.

1.3.1 The job performance requirements shall be accomplished in accordance with the requirements of the authority having jurisdiction and NFPA 1500, *Standard on Fire Department Occupational Safety and Health Program.*

1.3.2* It is not required for the job performance requirements to be mastered in the order they appear. The authority having jurisdiction shall establish instructional priority and the training program content to prepare individuals to meet the job performance requirements of this standard.

1.3.3* Performance of each requirement of this standard shall be evaluated by individuals approved by the authority having jurisdiction.

1.3.4 The entrance requirements in Chapter 4 of this standard shall be met prior to beginning training at the Fire Fighter I level.

1.3.5* Prior to being certified at the Fire Fighter I level, the fire fighter candidate shall meet the general knowledge and skills requirements and the job performance requirements of Chapter 5.

1.3.6 Prior to being certified at the Fire Fighter II level, the Fire Fighter I shall meet the general knowledge and skills requirements and the job performance requirements of Chapter 6.

1.3.7 Wherever in this standard the terms *rules, regulations, procedures, supplies, apparatus,* or *equipment* are referred to, it is implied that they are those of the authority having jurisdiction.

1.4 Units. In this standard, values for measurement are followed by an equivalent in SI units, but only the first stated value shall be regarded as the requirement. Equivalent values in SI units shall not be considered as the requirement, as these values can be approximate. (*See Table 1.4.*)

Table 1.4 SI Conversions

Quantity	*U.S. Unit/Symbol*	*SI Unit/Symbol*	*Conversion Factor*
Length	inch (in)	millimeter (mm)	1 in = 25.4 mm
	foot (ft)	meter (m)	1 ft = 0.305 m
Area	square foot (ft^2)	square meter (m^2)	1 ft^2 = 0.0929 m^2

Chapter 2 Referenced Publications

2.1 General. The documents or portions thereof listed in this chapter are referenced within this standard and shall be considered part of the requirements of this document.

2.2 NFPA Publications. National Fire Protection Association, 1 Batterymarch Park, P.O. Box 9101, Quincy, MA 02269-9101.

NFPA 472, *Standard for Professional Competence of Responders to Hazardous Materials Incidents,* 2002 edition.

NFPA 1500, *Standard on Fire Department Occupational Safety and Health Program,* 2002 edition.

NFPA 1582, *Standard on Medical Requirements for Fire Fighters and Information for Fire Department Physicians,* 2000 edition.

2.3 Other Publications. (Reserved)

Chapter 3 Definitions

3.1* General. The definitions contained in this chapter shall apply to the terms used in this standard. Where terms are not included, common usage of the terms shall apply.

3.2 NFPA Official Definitions.

3.2.1* Approved. Acceptable to the authority having jurisdiction.

3.2.2* Authority Having Jurisdiction (AHJ). The organization, office, or individual responsible for approving equipment, materials, an installation, or a procedure.

3.2.3* Listed. Equipment, materials, or services included in a list published by an organization that is acceptable to the authority having jurisdiction and concerned with evaluation of products or services, that maintains periodic inspection of production of listed equipment or materials or periodic evaluation of services, and whose listing states that either the equipment, material, or service meets appropriate designated standards or has been tested and found suitable for a specified purpose.

3.3 General Definitions.

3.3.1 Fire Department. An organization providing rescue, fire suppression, and related activities. The term "fire department" shall include any public, governmental, private, industrial, or military organization engaging in this type of activity.

3.3.2 Fire Fighter Candidate. The person who has fulfilled the entrance requirements of Chapter 4 of this standard but has not met the job performance requirements for Fire Fighter I.

3.3.3 Fire Fighter I. The person, at the first level of progression as defined in Chapter 5, who has demonstrated the knowledge and skills to function as an integral member of a fire-fighting team under direct supervision in hazardous conditions.

3.3.4* Fire Fighter II. The person, at the second level of progression as defined in Chapter 6, who has demonstrated the skills and depth of knowledge to function under general supervision.

3.3.5 Job Performance Requirement (JPR). A statement that describes a specific job task, lists the items necessary to complete the task, and defines measurable or observable outcomes and evaluation areas for the specific task.

3.3.6 Personal Protective Clothing. The full complement of garments fire fighters are normally required to wear while on emergency scene including turnout coat, protective trousers, fire-fighting boots, fire-fighting gloves, a protective hood, and a helmet with eye protection.

3.3.7 Personal Protective Equipment. Consists of full personal protective clothing, plus a self-contained breathing apparatus (SCBA) and a personal alert safety system (PASS) device.

3.3.8 Procedure. The series of actions, conducted in an approved manner and sequence, designed to achieve an intended outcome.

3.3.9 Requisite Knowledge. Fundamental knowledge one must have in order to perform a specific task.

3.3.10 Requisite Skills. The essential skills one must have in order to perform a specific task.

3.3.11 Structural Fire Fighting. The activities of rescue, fire suppression, and property conservation in buildings, enclosed structures, aircraft interiors, vehicles, vessels, aircraft, or like properties that are involved in a fire or emergency situation. [**1500:**3.3]

3.3.12 Task. A specific job behavior or activity.

3.3.13 Team. Two or more individuals who have been assigned a common task and are in proximity to and in direct communications with each other, coordinate their activities as a work group, and support the safety of one another.

Chapter 4 Entrance Requirements

4.1 General. Prior to entering training to meet the requirements of Chapters 5 and 6 of this standard, the candidate shall meet the following requirements:

(1) Minimum educational requirements established by the authority having jurisdiction
(2) Age requirements established by the authority having jurisdiction
(3)* Medical requirements of NFPA 1582, *Standard on Medical Requirements for Fire Fighters and Information for Fire Department Physicians*

4.2 Fitness Requirements. Physical fitness requirements for entry-level personnel shall be developed and validated by the authority having jurisdiction.

4.3* Emergency Medical Care. Minimum emergency medical care performance capabilities for entry level personnel shall be developed and validated by the authority having jurisdiction to include infection control, CPR, bleeding control, and shock management.

Chapter 5 Fire Fighter I

5.1 General.

5.1.1 For certification at Level I, the fire fighter candidate shall meet the general knowledge requirements in 5.1.1.1, the general skill requirements in 5.1.1.2, and the job performance requirements defined in Sections 5.2 through 5.5 of this standard and the requirements defined in Chapter 4, Competencies for the First Responder at the Awareness Level, of NFPA 472, *Standard for Professional Competence of Responders to Hazardous Materials Incidents.*

5.1.1.1 General Knowledge Requirements. The organization of the fire department; the role of the Fire Fighter I in the organization; the mission of fire service; the fire department's standard operating procedures and rules and regulations as they apply to the Fire Fighter I; the role of other agencies as they relate to the fire department; aspects of the fire department's member assistance program; the critical aspects of NFPA 1500, *Standard on Fire Department Occupational Safety and Health Program,* as they apply to the Fire Fighter I; knot types and usage; the difference between life safety and utility rope; reasons for placing rope out of service; the types of knots to use for given tools, ropes, or situations; hoisting methods for tools and equipment; and using rope to support response activities.

5.1.1.2 General Skill Requirements. The ability to don personal protective clothing within one minute; doff personal protective clothing and prepare for reuse; hoist tools and equipment using ropes and the correct knot; tie a bowline, clove hitch, figure eight on a bight, half hitch, becket or sheet bend, and safety knots; and locate information in departmental documents and standard or code materials.

5.2 Fire Department Communications. This duty involves initiating responses, receiving telephone calls, and using fire department communications equipment to correctly relay verbal or written information, according to the following job performance requirements.

5.2.1* Initiate the response to a reported emergency, given the report of an emergency, fire department standard operating procedures, and communications equipment, so that all necessary information is obtained, communications equipment is operated correctly, and the information is promptly and accurately relayed to the dispatch center.

(A) *Requisite Knowledge:* Procedures for reporting an emergency, departmental standard operating procedures for taking and receiving alarms, radio codes or procedures, and information needs of dispatch center.

(B) *Requisite Skills:* The ability to operate fire department communications equipment, relay information, and record information.

5.2.2 Receive a business or personal telephone call, given a fire department business phone, so that procedures for answering the phone are used and the caller's information is relayed.

(A) *Requisite Knowledge:* Fire department procedures for answering nonemergency telephone calls.

(B) *Requisite Skills:* The ability to operate fire station telephone and intercom equipment.

5.2.3 Transmit and receive messages via the fire department radio, given a fire department radio and operating procedures, so that the information is accurate, complete, clear, and relayed within the time established by the AHJ.

(A) *Requisite Knowledge:* Departmental radio procedures and etiquette for routine traffic, emergency traffic, and emergency evacuation signals.

(B) *Requisite Skills:* The ability to operate radio equipment and discriminate between routine and emergency traffic.

5.3 Fireground Operations. This duty involves performing activities necessary to ensure life safety, fire control, and property conservation, according to the following job performance requirements.

5.3.1* Use SCBA during emergency operations, given SCBA and other personal protective equipment, so that the SCBA is correctly donned and activated within one minute, the SCBA is correctly worn, controlled breathing techniques are used, emergency procedures are enacted if the SCBA fails, all low-air warnings are recognized, respiratory protection is not intentionally compromised, and hazardous areas are exited prior to air depletion.

(A) *Requisite Knowledge:* Conditions that require respiratory protection, uses and limitations of SCBA, components of SCBA, donning procedures, breathing techniques, indications for and emergency procedures used with SCBA, and physical requirements of the SCBA wearer.

(B) *Requisite Skills:* The ability to control breathing, replace SCBA air cylinders, use SCBA to exit through restricted passages, initiate and complete emergency procedures in the event of SCBA failure or air depletion, and complete donning procedures.

5.3.2* Respond on apparatus to an emergency scene, given personal protective clothing and other necessary personal protective equipment, so that the apparatus is correctly mounted and dismounted, seat belts are used while the vehicle is in motion, and other personal protective equipment is correctly used.

(A) *Requisite Knowledge:* Mounting and dismounting procedures for riding fire apparatus; hazards and ways to avoid hazards associated with riding apparatus; prohibited practices; types of department personal protective equipment and the means for usage.

(B) *Requisite Skills:* The ability to use each piece of provided safety equipment.

5.3.3* Operate in established work areas at emergency scenes, given protective equipment, traffic and scene control devices, structure fire and roadway emergency scenes, traffic hazards and downed electrical wires, so that procedures are followed, protective equipment is worn, protected work areas are established as directed using traffic and scene control devices, and the fire fighter performs assigned tasks only in established, protected work areas.

(A) *Requisite Knowledge:* Potential hazards involved in operating on emergency scenes including vehicle traffic, utilities, and environmental conditions; proper procedures for dismounting apparatus in traffic; procedures for safe operation at emergency scenes; and the protective equipment available for members' safety on emergency scenes and work zone designations.

(B) *Requisite Skills:* The ability to use PPC, the deployment of traffic and scene control devices, dismount apparatus and operate in the protected work areas as directed.

5.3.4* Force entry into a structure, given personal protective equipment, tools, and an assignment, so that the tools are used as designed, the barrier is removed, and the opening is in a safe condition and ready for entry.

(A) *Requisite Knowledge:* Basic construction of typical doors, windows, and walls within the department's community or service area; operation of doors, windows, and locks; and the dangers associated with forcing entry through doors, windows, and walls.

(B) *Requisite Skills:* The ability to transport and operate hand and power tools and to force entry through doors, windows, and walls using assorted methods and tools.

5.3.5* Exit a hazardous area as a team, given vision-obscured conditions, so that a safe haven is found before exhausting the air supply, others are not endangered, and the team integrity is maintained.

(A) *Requisite Knowledge:* Personnel accountability systems, communication procedures, emergency evacuation methods, what constitutes a safe haven, elements that create or indicate a hazard, and emergency procedures for loss of air supply.

(B) *Requisite Skills:* The ability to operate as a team member in vision-obscured conditions, locate and follow a guideline, conserve air supply, and evaluate areas for hazards and identify a safe haven.

5.3.6* Set up ground ladders, given single and extension ladders, an assignment, and team members if needed, so that hazards are assessed, the ladder is stable, the angle is correct for climbing, extension ladders are extended to the necessary height with the fly locked, the top is placed against a reliable structural component, and the assignment is accomplished.

(A) *Requisite Knowledge:* Parts of a ladder, hazards associated with setting up ladders, what constitutes a stable foundation for ladder placement, different angles for various tasks, safety limits to the degree of angulation, and what constitutes a reliable structural component for top placement.

(B) *Requisite Skills:* The ability to carry ladders, raise ladders, extend ladders and lock flies, determine that a wall and roof will support the ladder, judge extension ladder height requirements, and place the ladder to avoid obvious hazards.

5.3.7* Attack a passenger vehicle fire operating as a member of a team, given personal protective equipment, attack line, and hand tools, so that hazards are avoided, leaking flammable liquids are identified and controlled, protection from flash fires is maintained, all vehicle compartments are overhauled, and the fire is extinguished.

(A) *Requisite Knowledge:* Principles of fire streams as they relate to fighting automobile fires; precautions to be followed when advancing hose lines toward an automobile; observable results that a fire stream has been properly applied; identifying alternative fuels and the hazards associated with them; dangerous conditions created during an automobile fire; common types of accidents or injuries related to fighting automobile fires and how to avoid them; how to access locked passenger, trunk, and engine compartments; and methods for overhauling an automobile.

(B) *Requisite Skills:* The ability to identify automobile fuel type; assess and control fuel leaks; open, close, and adjust the flow and pattern on nozzles; apply water for maximum effectiveness while maintaining flash fire protection; advance 1½-in. (38-mm) or larger diameter attack lines; and expose hidden fires by opening all automobile compartments.

5.3.8* Extinguish fires in exterior Class A materials, given fires in stacked or piled and small unattached structures or storage containers that can be fought from the exterior, attack lines, hand tools and master stream devices, and an assignment, so that exposures are protected, the spread of fire is stopped, collapse hazards are avoided, water application is effective, the fire is extinguished, and signs of the origin area(s) and arson are preserved.

(A) *Requisite Knowledge:* Types of attack lines and water streams appropriate for attacking stacked, piled materials and outdoor fires; dangers—such as collapse—associated with stacked and piled materials; various extinguishing agents and their effect on different material configurations; tools and methods to use in breaking up various types of materials; the difficulties related to complete extinguishment of stacked and piled materials; water application methods for exposure protection and fire extinguishment; dangers such as exposure to toxic or hazardous materials associated with storage building and container fires; obvious signs of origin and cause; and techniques for the preservation of fire cause evidence.

(B) *Requisite Skills:* The ability to recognize inherent hazards related to the material's configuration, operate handlines or master streams, break up material using hand tools and water streams, evaluate for complete extinguishment, operate hose lines and other water application devices, evaluate and modify water application for maximum penetration, search for and expose hidden fires, assess patterns for origin determination, and evaluate for complete extinguishment.

5.3.9 Conduct a search and rescue in a structure operating as a member of a team, given an assignment, obscured vision conditions, personal protective equipment, a flashlight, forcible entry tools, hose lines, and ladders when necessary, so that ladders are correctly placed when used, all assigned areas are searched, all victims are located and removed, team integrity is maintained, and team members' safety—including respiratory protection—is not compromised.

(A) *Requisite Knowledge:* Use of forcible entry tools during rescue operations, ladder operations for rescue, psychological effects of operating in obscured conditions and ways to manage them, methods to determine if an area is tenable, primary and secondary search techniques, team members' roles and goals, methods to use and indicators of finding victims, victim removal methods (including various carries), and considerations related to respiratory protection.

(B) *Requisite Skills:* The ability to use SCBA to exit through restricted passages, set up and use different types of ladders for various types of rescue operations, rescue a fire fighter with functioning respiratory protection, rescue a fire fighter whose respiratory protection is not functioning, rescue a person who has no respiratory protection, and assess areas to determine tenability.

5.3.10* Attack an interior structure fire operating as a member of a team, given an attack line, ladders when needed, personal protective equipment, tools, and an assignment, so that team integrity is maintained, the attack line is deployed for advancement, ladders are correctly placed when used, access is gained into the fire area, effective water application practices are used, the fire is approached correctly, attack techniques facilitate suppression given the level of the fire, hidden fires are located and controlled, the correct body posture is maintained, hazards are recognized and managed, and the fire is brought under control.

(A) *Requisite Knowledge:* Principles of fire streams; types, design, operation, nozzle pressure effects, and flow capabilities of nozzles; precautions to be followed when advancing hose lines to a fire; observable results that a fire stream has been properly applied; dangerous building conditions created by fire; principles of exposure protection; potential long-term consequences of exposure to products of combustion; physical states of matter in which fuels are found; common types of accidents or injuries and their causes; and the application of each size and type of attack line, the role of the backup team in fire attack situations, attack and control techniques for grade level and above and below grade levels, and exposing hidden fires.

(B) *Requisite Skills:* The ability to prevent water hammers when shutting down nozzles; open, close, and adjust nozzle flow and patterns; apply water using direct, indirect, and combination attacks; advance charged and uncharged 1½-in. (38-mm) diameter or larger hose lines up ladders and up and down interior and exterior stairways; extend hose lines; replace burst hose sections; operate charged hose lines of 1½-in. (38-mm) diameter or larger while secured to a ground ladder; couple and uncouple various handline connections; carry hose; attack fires at grade level and above and below grade levels; and locate and suppress interior wall and subfloor fires.

5.3.11 Perform horizontal ventilation on a structure operating as part of a team, given an assignment, personal protective equipment, ventilation tools, equipment, and ladders, so that the ventilation openings are free of obstructions, tools are used as designed, ladders are correctly placed, ventilation devices are correctly placed, and the structure is cleared of smoke.

(A) *Requisite Knowledge:* The principles, advantages, limitations, and effects of horizontal, mechanical, and hydraulic ventilation; safety considerations when venting a structure; fire behavior in a structure; the products of combustion found in a structure fire; the signs, causes, effects, and prevention of backdrafts; and the relationship of oxygen concentration to life safety and fire growth.

(B) *Requisite Skills:* The ability to transport and operate ventilation tools and equipment and ladders and to use safe procedures for breaking window and door glass and removing obstructions.

5.3.12 Perform vertical ventilation on a structure as part of a team, given an assignment, personal protective equipment, ground and roof ladders, and tools, so that ladders are positioned for ventilation, a specified opening is created, all ventilation barriers are removed, structural integrity is not compromised, products of combustion are released from the structure, and the team retreats from the area when ventilation is accomplished.

(A) *Requisite Knowledge:* The methods of heat transfer; the principles of thermal layering within a structure on fire; the techniques and safety precautions for venting flat roofs, pitched roofs, and basements; basic indicators of potential collapse or roof failure; the effects of construction type and elapsed time under fire conditions on structural integrity; and the advantages and disadvantages of vertical and trench/strip ventilation.

(B) *Requisite Skills:* The ability to transport and operate ventilation tools and equipment; hoist ventilation tools to a roof; cut roofing and flooring materials to vent flat roofs, pitched roofs, and basements; sound a roof for integrity; clear an opening with hand tools; select, carry, deploy, and secure ground ladders for ventilation activities; deploy roof ladders on pitched roofs while secured to a ground ladder; and carry ventilation-related tools and equipment while ascending and descending ladders.

5.3.13 Overhaul a fire scene, given personal protective equipment, attack line, hand tools, a flashlight, and an assignment, so that structural integrity is not compromised, all hidden fires are discovered, fire cause evidence is preserved, and the fire is extinguished.

(A) *Requisite Knowledge:* Types of fire attack lines and water application devices most effective for overhaul, water application methods for extinguishment that limit water damage, types of tools and methods used to expose hidden fire, dangers associated with overhaul, obvious signs of area of origin or signs of arson, and reasons for protection of fire scene.

(B) *Requisite Skills:* The ability to deploy and operate an attack line; remove flooring, ceiling, and wall components to expose void spaces without compromising structural integrity; apply water for maximum effectiveness; expose and extinguish hidden fires in walls, ceilings, and subfloor spaces; recognize and preserve obvious signs of area of origin and arson; and evaluate for complete extinguishment.

5.3.14 Conserve property as a member of a team, given salvage tools and equipment and an assignment, so that the building and its contents are protected from further damage.

(A) *Requisite Knowledge:* The purpose of property conservation and its value to the public, methods used to protect property, types of and uses for salvage covers, operations at properties protected with automatic sprinklers, how to stop the flow of water from an automatic sprinkler head, identification of the main control valve on an automatic sprinkler system, and forcible entry issues related to salvage.

(B) *Requisite Skills:* The ability to cluster furniture; deploy covering materials; roll and fold salvage covers for reuse; construct water chutes and catch-alls; remove water; cover building openings, including doors, windows, floor openings, and roof openings; separate, remove, and relocate charred material to a safe location while protecting the area of origin for cause determination; stop the flow of water from a sprinkler with sprinkler wedges or stoppers; and operate a main control valve on an automatic sprinkler system.

5.3.15* Connect a fire department pumper to a water supply as a member of a team, given supply or intake hose, hose tools, and a fire hydrant or static water source, so that connections are tight and water flow is unobstructed.

(A) *Requisite Knowledge:* Loading and off-loading procedures for mobile water supply apparatus; fire hydrant operation; and suitable static water supply sources, procedures, and protocol for connecting to various water sources.

(B) *Requisite Skills:* The ability to hand lay a supply hose, connect and place hard suction hose for drafting operations, deploy portable water tanks as well as the equipment necessary to transfer water between and draft from them, make hydrant-to-pumper hose connections for forward and reverse lays, connect supply hose to a hydrant, and fully open and close the hydrant.

5.3.16* Extinguish incipient Class A, Class B, and Class C fires, given a selection of portable fire extinguishers, so that the correct extinguisher is chosen, the fire is completely extinguished, and correct extinguisher-handling techniques are followed.

(A) *Requisite Knowledge:* The classifications of fire; the types of, rating systems for, and risks associated with each class of fire; and the operating methods of, and limitations of portable extinguishers.

(B) *Requisite Skills:* The ability to operate portable fire extinguishers, approach fire with portable fire extinguishers, select an appropriate extinguisher based on the size and type of fire, and safely carry portable fire extinguishers.

5.3.17 Illuminate the emergency scene, given fire service electrical equipment and an assignment, so that designated areas are illuminated and all equipment is operated within the manufacturer's listed safety precautions.

(A) *Requisite Knowledge:* Safety principles and practices, power supply capacity and limitations, and light deployment methods.

(B) *Requisite Skills:* The ability to operate department power supply and lighting equipment, deploy cords and connectors, reset ground-fault interrupter (GFI) devices, and locate lights for best effect.

5.3.18 Turn off building utilities, given tools and an assignment, so that the assignment is safely completed.

(A) *Requisite Knowledge:* Properties, principles, and safety concerns for electricity, gas, and water systems; utility disconnect methods and associated dangers; and use of required safety equipment.

(B) *Requisite Skills:* The ability to identify utility control devices, operate control valves or switches, and assess for related hazards.

5.3.19* Combat a ground cover fire operating as a member of a team, given protective clothing, SCBA if needed, hose lines, extinguishers or hand tools, and an assignment, so that threats to property are reported, threats to personal safety are recognized, retreat is quickly accomplished when warranted, and the assignment is completed.

(A) *Requisite Knowledge:* Types of ground cover fires, parts of ground cover fires, methods to contain or suppress, and safety principles and practices.

(B) *Requisite Skills:* The ability to determine exposure threats based on fire spread potential, protect exposures, construct a fire line or extinguish with hand tools, maintain integrity of established fire lines, and suppress ground cover fires using water.

5.4 Rescue Operations. This duty involves no requirements for Fire Fighter I.

5.5 Prevention, Preparedness, and Maintenance. This duty involves performing activities that reduce the loss of life and property due to fire through hazard identification, inspection, education, and response readiness, according to the following job performance requirements.

5.5.1 Perform a fire safety survey in a private dwelling, given survey forms and procedures, so that fire and life-safety hazards are identified, recommendations for their correction are made to the occupant, and unresolved issues are referred to the proper authority.

(A) *Requisite Knowledge:* Organizational policy and procedures, common causes of fire and their prevention, the importance of a fire safety survey and public fire education programs to fire department public relations and the community, and referral procedures.

(B) *Requisite Skills:* The ability to complete forms, recognize hazards, match findings to preapproved recommendations, and effectively communicate findings to occupants or referrals.

5.5.2* Present fire safety information to station visitors or small groups, given prepared materials, so that all information is presented, the information is accurate, and questions are answered or referred.

(A) *Requisite Knowledge:* Parts of informational materials and how to use them, basic presentation skills, and departmental standard operating procedures for giving fire station tours.

(B) *Requisite Skills:* The ability to document presentations and to use prepared materials.

5.5.3 Clean and check ladders, ventilation equipment, self-contained breathing apparatus (SCBA), ropes, salvage equipment, and hand tools, given cleaning tools, cleaning supplies, and an assignment, so that equipment is clean and maintained according to manufacturer's or departmental guidelines, maintenance is recorded, and equipment is placed in a ready state or reported otherwise.

(A) *Requisite Knowledge:* Types of cleaning methods for various tools and equipment, correct use of cleaning solvents, and manufacturer's or departmental guidelines for cleaning equipment and tools.

(B) *Requisite Skills:* The ability to select correct tools for various parts and pieces of equipment, follow guidelines, and complete recording and reporting procedures.

5.5.4 Clean, inspect, and return fire hose to service, given washing equipment, water, detergent, tools, and replacement gaskets, so that damage is noted and corrected, the hose is clean, and the equipment is placed in a ready state for service.

(A) *Requisite Knowledge:* Departmental procedures for noting a defective hose and removing it from service, cleaning methods, and hose rolls and loads.

(B) *Requisite Skills:* The ability to clean different types of hose; operate hose washing and drying equipment; mark defective hose; and replace coupling gaskets, roll hose, and reload hose.

Chapter 6 Fire Fighter II

6.1 General.

6.1.1 For certification at Level II, the Fire Fighter I shall meet the general knowledge requirements in 6.1.1.1, the general skill requirements in 6.1.1.2, and the job performance requirements defined in Sections 6.2 through 6.5 of this standard and the requirements defined in Chapter 5, Competencies for the First Responder at the Operational Level, of NFPA 472, *Standard for Professional Competence of Responders to Hazardous Materials Incidents.*

6.1.1.1 General Knowledge Requirements. Responsibilities of the Fire Fighter II in assuming and transferring command within an incident management system, performing assigned duties in conformance with applicable NFPA and other safety regulations and authority having jurisdiction procedures, and the role of a Fire Fighter II within the organization.

6.1.1.2 General Skill Requirements. The ability to determine the need for command, organize and coordinate an incident management system until command is transferred, and function within an assigned role in the incident management system.

6.2 Fire Department Communications. This duty involves performing activities related to initiating and reporting responses, according to the following job performance requirements.

6.2.1 Complete a basic incident report, given the report forms, guidelines, and information, so that all pertinent information is recorded, the information is accurate, and the report is complete.

(A) *Requisite Knowledge:* Content requirements for basic incident reports, the purpose and usefulness of accurate reports, consequences of inaccurate reports, how to obtain necessary information, and required coding procedures.

(B) *Requisite Skills:* The ability to determine necessary codes, proof reports, and operate fire department computers or other equipment necessary to complete reports.

6.2.2* Communicate the need for team assistance, given fire department communications equipment, standard operating procedures (SOPs), and a team, so that the supervisor is consistently informed of team needs, departmental SOPs are followed, and the assignment is accomplished safely.

(A) *Requisite Knowledge:* SOPs for alarm assignments and fire department radio communication procedures.

(B) *Requisite Skills:* The ability to operate fire department communications equipment.

6.3 Fireground Operations. This duty involves performing activities necessary to insure life safety, fire control, and property conservation, according to the following job performance requirements.

6.3.1* Extinguish an ignitable liquid fire, operating as a member of a team, given an assignment, an attack line, personal protective equipment, a foam proportioning device, a nozzle, foam concentrates, and a water supply, so that the correct type of foam concentrate is selected for the given fuel and conditions, a properly proportioned foam stream is applied to the surface of the fuel to create and maintain a foam blanket, fire is extinguished, reignition is prevented, team protection is maintained with a foam stream, and the hazard is faced until retreat to safe haven is reached.

(A) *Requisite Knowledge:* Methods by which foam prevents or controls a hazard; principles by which foam is generated; causes for poor foam generation and corrective measures; difference between hydrocarbon and polar solvent fuels and the concentrates that work on each; the characteristics, uses, and limitations of fire-fighting foams; the advantages and disadvantages of using fog nozzles versus foam nozzles for foam application; foam stream application techniques; hazards associated with foam usage; and methods to reduce or avoid hazards.

(B) *Requisite Skills:* The ability to prepare a foam concentrate supply for use, assemble foam stream components, master various foam application techniques, and approach and retreat from spills as part of a coordinated team.

6.3.2* Coordinate an interior attack line for team's accomplishment of an assignment in a structure fire, given attack lines, personnel, personal protective equipment, and tools, so that crew integrity is established; attack techniques are selected for the given level of the fire (for example, attic, grade level, upper levels, or basement); attack techniques are communicated to the attack teams; constant team coordination is maintained; fire growth and development is continuously evaluated; search, rescue, and ventilation requirements are communicated or managed; hazards are reported to the attack teams; and incident command is apprised of changing conditions.

(A) *Requisite Knowledge:* Selection of the nozzle and hose for fire attack given different fire situations; selection of adapters and appliances to be used for specific fire ground situations; dangerous building conditions created by fire and fire suppression activities; indicators of building collapse; the effects of fire and fire suppression activities on wood, masonry (brick, block, stone), cast iron, steel, reinforced concrete, gypsum wall board, glass, and plaster on lath; search and rescue and ventilation procedures; indicators of structural instability; suppression approaches and practices for various types of structural fires; and the association between specific tools and special forcible entry needs.

(B) *Requisite Skills:* The ability to assemble a team, choose attack techniques for various levels of a fire (e.g., attic, grade level, upper levels, or basement), evaluate and forecast a fire's growth and development, select tools for forcible entry, incorporate search and rescue procedures and ventilation procedures in the completion of the attack team efforts, and determine developing hazardous building or fire conditions.

6.3.3* Control a flammable gas cylinder fire operating as a member of a team, given an assignment, a cylinder outside of a structure, an attack line, personal protective equipment, and tools, so that crew integrity is maintained, contents are identified, safe havens are identified prior to advancing, open valves are closed, flames are not extinguished unless the leaking gas is eliminated, the cylinder is cooled, cylinder integrity is evaluated, hazardous conditions are recognized and acted upon, and the cylinder is faced during approach and retreat.

(A) *Requisite Knowledge:* Characteristics of pressurized flammable gases, elements of a gas cylinder, effects of heat and pressure on closed cylinders, boiling liquid expanding vapor explosion (BLEVE) signs and effects, methods for identifying contents, how to identify safe havens before approaching flammable gas cylinder fires, water stream usage and demands for pressurized cylinder fires, what to do if the fire is prematurely extinguished, valve types and their operation, alternative actions related to various hazards and when to retreat.

(B) *Requisite Skills:* The ability to execute effective advances and retreats, apply various techniques for water application, assess cylinder integrity and changing cylinder conditions, operate control valves, choose effective procedures when conditions change.

6.3.4* Protect evidence of fire cause and origin, given a flashlight and overhaul tools, so that the evidence is noted and protected from further disturbance until investigators can arrive on the scene.

(A) *Requisite Knowledge:* Methods to assess origin and cause; types of evidence; means to protect various types of evidence; the role and relationship of Fire Fighter IIs, criminal investigators, and insurance investigators in fire investigations; and the effects and problems associated with removing property or evidence from the scene.

(B) *Requisite Skills:* The ability to locate the fire's origin area, recognize possible causes, and protect the evidence.

6.4 Rescue Operations. This duty involves performing activities related to accessing and disentangling victims from motor vehicle accidents and helping special rescue teams, according to the following job performance requirements.

6.4.1* Extricate a victim entrapped in a motor vehicle as part of a team, given stabilization and extrication tools, so that the vehicle is stabilized, the victim disentangled without further injury, and hazards are managed.

(A) *Requisite Knowledge:* The fire department's role at a vehicle accident, points of strength and weakness in auto body construction, dangers associated with vehicle components and systems, the uses and limitations of hand and power extrication equipment, and safety procedures when using various types of extrication equipment.

(B) *Requisite Skills:* The ability to operate hand and power tools used for forcible entry and rescue as designed; use cribbing and shoring material; and choose and apply appropriate techniques for moving or removing vehicle roofs, doors, windshields, windows, steering wheels or columns, and the dashboard.

6.4.2* Assist rescue operation teams, given standard operating procedures, necessary rescue equipment, and an assignment, so that procedures are followed, rescue items are recognized and retrieved in the time as prescribed by the AHJ, and the assignment is completed.

(A) *Requisite Knowledge:* The fire fighter's role at a special rescue operation, the hazards associated with special rescue operations, types and uses for rescue tools, and rescue practices and goals.

(B) *Requisite Skills:* The ability to identify and retrieve various types of rescue tools, establish public barriers, and assist rescue teams as a member of the team when assigned.

6.5 Prevention, Preparedness, and Maintenance. This duty involves performing activities related to reducing the loss of life and property due to fire through hazard identification, inspection, and response readiness, according to the following job performance requirements.

6.5.1* Prepare a preincident survey, given forms, necessary tools, and an assignment, so that all required occupancy information is recorded, items of concern are noted, and accurate sketches or diagrams are prepared.

(A) *Requisite Knowledge:* The sources of water supply for fire protection; the fundamentals of fire suppression and detection systems; common symbols used in diagramming construction features, utilities, hazards, and fire protection systems; departmental requirements for a preincident survey and form completion; and the importance of accurate diagrams.

(B) *Requisite Skills:* The ability to identify the components of fire suppression and detection systems; sketch the site, buildings, and special features; detect hazards and special considerations to include in the preincident sketch; and complete all related departmental forms.

6.5.2 Maintain power plants, power tools, and lighting equipment, given tools and manufacturers' instructions, so that equipment is clean and maintained according to manufacturer and departmental guidelines, maintenance is recorded, and equipment is placed in a ready state or reported otherwise.

(A) *Requisite Knowledge:* Types of cleaning methods, correct use of cleaning solvents, manufacturer and departmental guidelines for maintaining equipment and its documentation, and problem-reporting practices.

(B) *Requisite Skills:* The ability to select correct tools; follow guidelines; complete recording and reporting procedures; and operate power plants, power tools, and lighting equipment.

6.5.3 Perform an annual service test on fire hose, given a pump, a marking device, pressure gauges, a timer, record sheets, and related equipment, so that procedures are followed, the condition of the hose is evaluated, any damaged hose is removed from service, and the results are recorded.

(A)* *Requisite Knowledge:* Procedures for safely conducting hose service testing, indicators that dictate any hose be removed from service, and recording procedures for hose test results.

(B) *Requisite Skills:* The ability to operate hose testing equipment and nozzles and to record results.

6.5.4* Test the operability of and flow from a fire hydrant, given a Pitot tube, pressure gauge, and other necessary tools, so that the readiness of the hydrant is assured and the flow of water from the hydrant can be calculated and recorded.

(A) *Requisite Knowledge:* How water flow is reduced by hydrant obstructions; direction of hydrant outlets to suitability of use; the effect of mechanical damage, rust, corrosion, failure to open the hydrant fully, and susceptibility to freezing; and the meaning of the terms *static, residual,* and *flow pressure.*

(B) *Requisite Skills:* The ability to operate a pressurized hydrant, use a Pitot tube and pressure gauges, detect damage, and record results of test.

Annex A Explanatory Material

Annex A is not a part of the requirements of this NFPA document but is included for informational purposes only. This annex contains explanatory material, numbered to correspond with the applicable text paragraphs.

A.1.3.2 See Annex B for additional information regarding the use of job performance requirements for training and evaluation.

A.1.3.3 It is recommended, where practical, that evaluators be individuals who were not directly involved as instructors for the requirement being evaluated.

A.1.3.5 Many jurisdictions choose to deliver Fire Fighter I training in modules that allow personnel to be trained in certain fire fighter tasks and to perform limited duties under direct supervision prior to meeting the complete requirements for Fire Fighter I certification.

A.3.1 Definitions of action verbs used in the job performance requirements in this document are based on the first definition of the word found in *Webster's Third New International Dictionary of the English Language.*

A.3.2.1 Approved. The National Fire Protection Association does not approve, inspect, or certify any installations, procedures, equipment, or materials; nor does it approve or evaluate testing laboratories. In determining the acceptability of installations, procedures, equipment, or materials, the authority having jurisdiction may base acceptance on compliance with NFPA or other appropriate standards. In the absence of such standards, said authority may require evidence of proper installation, procedure, or use. The authority having jurisdiction may also refer to the listings or labeling practices of an organization that is concerned with product evaluations and is thus in a position to determine compliance with appropriate standards for the current production of listed items.

A.3.2.2 Authority Having Jurisdiction (AHJ). The phrase "authority having jurisdiction," or its acronym AHJ, is used in NFPA documents in a broad manner, since jurisdictions and approval agencies vary, as do their responsibilities. Where public safety is primary, the authority having jurisdiction may be a federal, state, local, or other regional department or individual such as a fire chief; fire marshal; chief of a fire prevention bureau, labor department, or health department; building official; electrical inspector; or others having statutory authority. For insurance purposes, an insurance inspection department, rating bureau, or other insurance company representative may be the authority having jurisdiction. In many circumstances, the property owner or his or her designated agent assumes the role of the authority having jurisdiction; at government installations, the commanding officer or departmental official may be the authority having jurisdiction.

A.3.2.3 Listed. The means for identifying listed equipment may vary for each organization concerned with product evaluation; some organizations do not recognize equipment as listed unless it is also labeled. The authority having jurisdiction should utilize the system employed by the listing organization to identify a listed product.

A.3.3.4 Fire Fighter II. This person will function as an integral member of a team of equally or less experienced fire fighters to accomplish a series of tasks. When engaged in hazardous activities, the Fire Fighter II maintains direct communications with a supervisor.

A.4.1(3) The candidate should meet the requirements of NFPA 1582, *Standard on Medical Requirements for Fire Fighters and Information for Fire Department Physicians,* within a reasonable period of time prior to entering into training or testing for Fire Fighter I to ensure his or her ability to safely perform the required tasks.

A.4.3 Programs such as the Department of Transportation First Responder and American Red Cross curricula offer models that can be followed.

A.5.2.1 The Fire Fighter I should be able to receive and accurately process information received at the station. Fire fighters used as telecommunicators (dispatchers) should meet the requirements of NFPA 1061, *Standard for Professional Qualifications for Public Safety Telecommunicator,* for qualification standards and job performance requirements.

A.5.3.1 The Fire Fighter I should already be wearing full protective clothing prior to the beginning of this SCBA-donning procedure. In addition to fully donning and activating the SCBA, the Fire Fighter I should also replace any personal protective clothing (i.e., gloves, protective hood, helmet, etc.) displaced during the donning procedure and activate the PASS device within the specified 1-minute time limit.

A.5.3.2 Other personal protective equipment might include hearing protection in cabs that have a noise level in excess of 90 dBa, eye protection for fire fighters riding in jump seats that are not fully enclosed, and SCBAs for those departments that require fire fighters to don SCBAs while en route to the emergency.

A.5.3.3 The safety of responders operating at an emergency scene is a key concern and one of the primary skills that the fire fighter must develop. Operations on roads and highways, on scenes where visibility is restricted, or where utilities may be unstable present a significant risk to the fire fighter as they dismount from apparatus and initiate emergency operations. Special protective equipment and constant attention to potential hazards is essential.

Fire fighters can be assigned to direct the movement of traffic at the scene or set up flare or cone lines either independently or in conjunction with law/traffic enforcement officers. A firefighter assigned to this duty (either briefly or until the incident is under control) should understand the proper techniques to control traffic and the appropriate use of protective clothing and signaling equipment.

A.5.3.4 The Fire Fighter I should be able to force entry through wood, glass, and metal doors that open in and out; overhead doors; and windows common to the community or service area.

A.5.3.5 When training exercises are intended to simulate emergency conditions, smoke-generating devices that do not create a hazard are required. Several accidents have occurred when smoke bombs or other smoke-generating devices that produce a toxic atmosphere have been used for training exercises. All exercises should be conducted in accordance with the requirements of NFPA 1404, *Standard for Fire Service Respiratory Protection Training.*

A.5.3.6 The fire fighter should be able to accomplish this task with each type and length of ground ladder carried by the department.

A.5.3.7 Passenger vehicles include automobiles, light trucks, and vans.

A.5.3.8 The Fire Fighter I should be able to extinguish fires in stacked or piled materials such as hay bales, pallets, lumber, piles of mulch, sawdust, other bulk Class A materials, or small unattached structures that are attacked from the exterior. The tactics for extinguishing each of these types of fires are similar enough to be included in one JPR.

Live fire evolutions should be conducted in accordance with the requirements of NFPA 1403, *Standard on Live Fire Training Evolutions.* It is further recommended that prior to involvement in live fire evolutions, the fire fighter demonstrate the use of SCBA in smoke and elevated temperature conditions.

In areas where environmental or other concerns restrict the use of Class A fuels for training evolutions, properly installed and monitored gas-fueled fire simulators might be substituted.

A.5.3.10 The Fire Fighter I should be proficient in the various attack approaches for room and contents fires at three different levels (at grade, above grade, and below grade). Maintenance of body posture in the standard refers to staying low during initial attack, protecting oneself from falling objects, and otherwise using common sense given the state of the fire's growth or suppression.

Live fire evolutions should be conducted in accordance with the requirements of NFPA 1403, *Standard on Live Fire Training Evolutions.* It is further recommended that prior to involvement in live fire evolutions, the fire fighter demonstrate the use of SCBA in smoke and elevated temperature conditions.

In areas where environmental or other concerns restrict the use of Class A fuels for training evolutions, properly installed and monitored gas-fueled fire simulators might be substituted.

A.5.3.15 Static water sources can include portable water tanks, ponds, creeks, and so forth.

A.5.3.16 The Fire Fighter I should be able to extinguish incipient Class A fires such as wastebaskets, small piles of pallets, wood, or hay;

Class B fires of approximately 9 ft^2 (0.84 m^2); and Class C fires where the electrical equipment is energized.

A.5.3.19 Protective clothing is not personal protective clothing as used throughout the rest of this document. Some jurisdictions provide fire fighters with different clothing for ground cover fires than is worn for structural fires. This clothing can be substituted for structural protective clothing in order to meet the intent of this job performance requirement.

A.5.5.2 The Fire Fighter I should be able to present basic information on how to (1) stop, drop, and roll when one's clothes are on fire; (2) crawl low in smoke; (3) perform escape planning; (4) alert others of an emergency; (5) call the fire department; and (6) properly place, test, and maintain residential smoke detectors. The Fire Fighter I is not expected to be an accomplished speaker or instructor.

A.6.2.2 The Fire Fighter II could be assigned to accomplish or coordinate tasks away from direct supervision. Many of these tasks could result in the need for additional or replacement personnel due to the ever-changing conditions on the scene of an emergency. The Fire Fighter II is expected to identify these needs and effectively communicate this information within an incident management system. Use of radio communication equipment necessitates that these communications be accurate and efficient.

A.6.3.1 The Fire Fighter II should be able to accomplish this task with each type of foam concentrate used by the jurisdiction. This could include the use of both Class A and B foam concentrates on appropriate fires. When using Class B foams to attack flammable or combustible liquid fires, the Fire Fighter II should extinguish a fire of at least 100 ft^2 (9 m^2). The Fire Fighter II is not expected to calculate application rates and densities. The intent of this JPR can be met in training through the use of training foam concentrates or gas-fired training props.

A.6.3.2 The Fire Fighter II should be able to coordinate the actions of the interior attack line team at common residential fires and small business fires in the fire department's district. Complex or large interior fire management should be left to the officers; however, this job performance requirement will facilitate the development of the Fire Fighter II towards effectively handling specific assignments within large fires.

Jurisdictions that use Fire Fighter IIs as acting company officers should comply with the requirements of NFPA 1021, *Standard for Fire Officer Professional Qualifications.*

A.6.3.3 Controlling flammable gas cylinder fires can be a very dangerous operation. The Fire Fighter II should act as a team member, under the direct supervision of an officer, during these operations.

A.6.3.4 The Fire Fighter II should be able to recognize important evidence as to a fire's cause and maintain the evidence so that further testing can be done without contamination or chain-of-custody problems. Evidence should be left in place (when possible, otherwise chain-of-custody must be established), not altered by improper handling, walking, and so forth, and not destroyed. Possible means to protect evidence is to avoid touching, protect with salvage covers during overhaul, or rope off the area where the evidence lies. The Fire Fighter II is not intended to be highly proficient at origin and cause determination.

Jurisdictions that use Fire Fighter IIs to determine origin and cause should comply with the requirements of NFPA 1021, *Standard for Fire Officer Professional Qualifications.*

A.6.4.1 In the context of this standard, the term *extricate* refers to those activities required to allow emergency medical personnel access to the victim, stabilization of the vehicle, the displacement or removal of vehicle components obstructing victim removal, and the protection of the victim and response personnel from hazards associated with motor vehicle accidents and the use of hand and power tools on a motor vehicle.

As persons performing extrication can be different from those performing medical functions, this standard does not address medical care of the victim. An awareness of the needs and responsibilities of emergency medical functions is recommended to allow for efficient coordination between the "extrication" team and the "medical" team.

A.6.4.2 The Fire Fighter II is not expected to be proficient in special rescue skills. The Fire Fighter II should be able to help special rescue teams in their efforts to safely manage structural collapses, trench collapses, cave and tunnel emergencies, water and ice emergencies, elevator and escalator emergencies, energized electrical line emergencies, and industrial accidents.

A.6.5.1 The Fire Fighter II should be able to compile information related to potential emergency incidents within their community for use by officers in the development of preincident plans. Jurisdictions that use Fire Fighter IIs to develop preincident plans should comply with the requirements of NFPA 1021, *Standard for Fire Officer Professional Qualifications.*

A.6.5.3(A) Procedures for conducting hose testing can be found in Chapter 5, Service Testing, of NFPA 1962, *Standard for the Care, Use, and Service Testing of Fire Hose Including Couplings and Nozzles.*

A.6.5.4 All fire fighters should be able to flow test a hydrant. While not all fire departments have hydrants in their jurisdiction, departments without hydrants in their jurisdiction can effectively train and evaluate a Fire Fighter II's flow testing skills by using hose streams.

Annex B Explanation of the Standard and Concepts of JPRs

This annex is not a part of the requirements of this NFPA document but is included for informational purposes only.

B.1 Explanation of the Standard and Concepts of Job Performance Requirements (JPRs). The primary benefit of establishing national professional qualification standards is to provide both public and private sectors with a framework of the job requirements for the fire service. Other benefits include enhancement of the profession, individual as well as organizational growth and development, and standardization of practices.

NFPA professional qualifications standards identify the minimum JPRs for specific fire service positions. The standards can be used for training design and evaluation, certification, measuring and critiquing on-the-job performance, defining hiring practices, and setting organizational policies, procedures, and goals. (Other applications are encouraged.)

Professional qualifications standards for a specific job are organized by major areas of responsibility defined as duties. For example, the fire fighter's duties might include fire suppression, rescue, and water supply; and the public fire educator's duties might include education, planning and development, and administration. Duties are major functional areas of responsibility within a job.

The professional qualifications standards are written as JPRs. JPRs describe the performance required for a specific job. JPRs are grouped according to the duties of a job. The complete list of JPRs for each duty defines what an individual must be able to do in order to successfully perform that duty. Together, the duties and their JPRs define the job parameters, that is, the standard as a whole is a description of a job.

B.2 Breaking Down the Components of a JPR. The JPR is the assembly of three critical components. *(See Table B.2.)* These components are as follows:

(1) Task that is to be performed
(2) Tools, equipment, or materials that must be provided to successfully complete the task
(3) Evaluation parameters and/or performance outcomes

Table B.2 Example of a JPR

(1) Task	(1) Ventilate a pitched roof
(2) Tools, equipment, or materials	(2) Given an ax, a pike pole, an extension ladder, and a roof ladder
(3) Evaluation parameters and performance outcomes	(3) So that a 4-ft × 4-ft hole is created; all ventilation barriers are removed; ladders are properly positioned for ventilation; ventilation holes are correctly placed; and smoke, heat, and combustion by-products are released from the structure

B.2.1 The Task to Be Performed. The first component is a concise, brief statement of what the person is supposed to do.

B.2.2 Tools, Equipment, or Materials that Must be Provided to Successfully Complete the Task. This component ensures that all individuals completing the task are given the same minimal tools, equipment, or materials when being evaluated. By listing these items, the performer and evaluator know what must be provided in order to complete the task.

B.2.3 Evaluation Parameters and/or Performance Outcomes. This component defines how well one must perform each task—for both the performer and the evaluator. The JPR guides performance towards successful completion by identifying evaluation parameters and/or performance outcomes. This portion of the JPR promotes consistency in evaluation by reducing the variables used to gauge performance.

In addition to these three components, the JPR contains requisite knowledge and skills. Just as the term *requisite* suggests, these are the necessary knowledge and skills one must have prior to being able to perform the task. Requisite knowledge and skills are the foundation for task performance.

Once the components and requisites are put together, the JPR might read as follows.

B.2.3.1 Example 1. The Fire Fighter I shall ventilate a pitched roof, given an ax, a pike pole, an extension ladder, and a roof ladder, so that a 4-ft × 4-ft hole is created, all ventilation barriers are removed, ladders are properly positioned for ventilation, and ventilation holes are correctly placed.

(A) *Requisite Knowledge:* Pitched roof construction, safety considerations with roof ventilation, the dangers associated with improper ventilation, knowledge of ventilation tools, the effects of ventilation on fire growth, smoke movement in structures, signs of backdraft, and the knowledge of vertical and forced ventilation.

(B) *Requisite Skills:* The ability to remove roof covering; properly initiate roof cuts; use the pike pole to clear ventilation barriers; use ax properly for sounding, cutting, and stripping; position ladders; and climb and position self on ladder.

B.2.3.2 Example 2. The Fire Investigator shall interpret burn patterns, given standard equipment and tools and some structural/content remains, so that each individual pattern is evaluated with respect to the burning characteristics of the material involved.

(A) *Requisite Knowledge:* Knowledge of fire development and the interrelationship of heat release rate, form, and ignitability of materials.

(B) *Requisite Skill:* The ability to interpret the effects of burning characteristics on different types of materials.

B.3 Examples of Potential Uses

B.3.1 Certification. JPRs can be used to establish the evaluation criteria for certification at a specific job level. When used for certification, evaluation must be based on the successful completion of JPRs.

First, the evaluator would verify the attainment of requisite knowledge and skills prior to JPR evaluation. Verification might be accomplished through documentation review or testing.

Next, the candidate would be evaluated on completing the JPRs. The candidate would perform the task and be evaluated based on the evaluation parameters, the performance outcomes, or both. This performance-based evaluation can be either practical (for psychomotor skills such as "ventilate a roof") or written (for cognitive skills such as "interpret burn patterns").

Note that psychomotor skills are those physical skills that can be demonstrated or observed. Cognitive skills (or mental skills) cannot be observed, but are rather evaluated on how one completes the task (process oriented) or the task outcome (product oriented).

Using Example 1, a practical performance-based evaluation would measure one's ability to "ventilate a pitched roof." The candidate passes this particular evaluation if the standard was met—that is, a 4-ft × 4-ft hole was created; all ventilation barriers were removed; ladders were properly positioned for ventilation; ventilation holes were correctly placed; and smoke, heat, and combustion by-products were released from the structure.

For Example 2, when evaluating the task "interpret burn patterns," the candidate could be given a written assessment in the form of a scenario, photographs, and drawings and then be asked to respond to specific written questions related to the JPR's evaluation parameters.

Remember, when evaluating performance, you must give the person the tools, equipment, or materials listed in the job performance requirements—for example, an ax, a pike pole, an extension ladder, and a roof ladder—before he or she can be properly evaluated.

B.3.2 Curriculum Development/Training Design and Evaluation. The statements contained in this document that refer to job performance were designed and written as JPRs. Although a resemblance to instructional objectives might be present, these statements should not be used in a teaching situation until after they have been modified for instructional use.

JPRs state the behaviors required to perform specific skill(s) on the job, as opposed to a learning situation. These statements should be converted into instructional objectives with behaviors, conditions, and standards that can be measured within the teaching/learning environment. A JPR that requires a fire fighter to "ventilate a pitched roof" should be converted into a measurable instructional objective for use when teaching the skill. *[See Figure B.3.2(a).]*

Using Example 1, a terminal instructional objective might read as follows.

The learner will ventilate a pitched roof, given a simulated roof, an ax, a pike pole, an extension ladder, and a roof ladder, so that 100 percent accuracy is attained on a skills checklist. (At a minimum, the skills checklist should include each of the measurement criteria from the job performance requirements.)

Figure B.3.2(b) is a sample checklist for use in evaluating this objective.

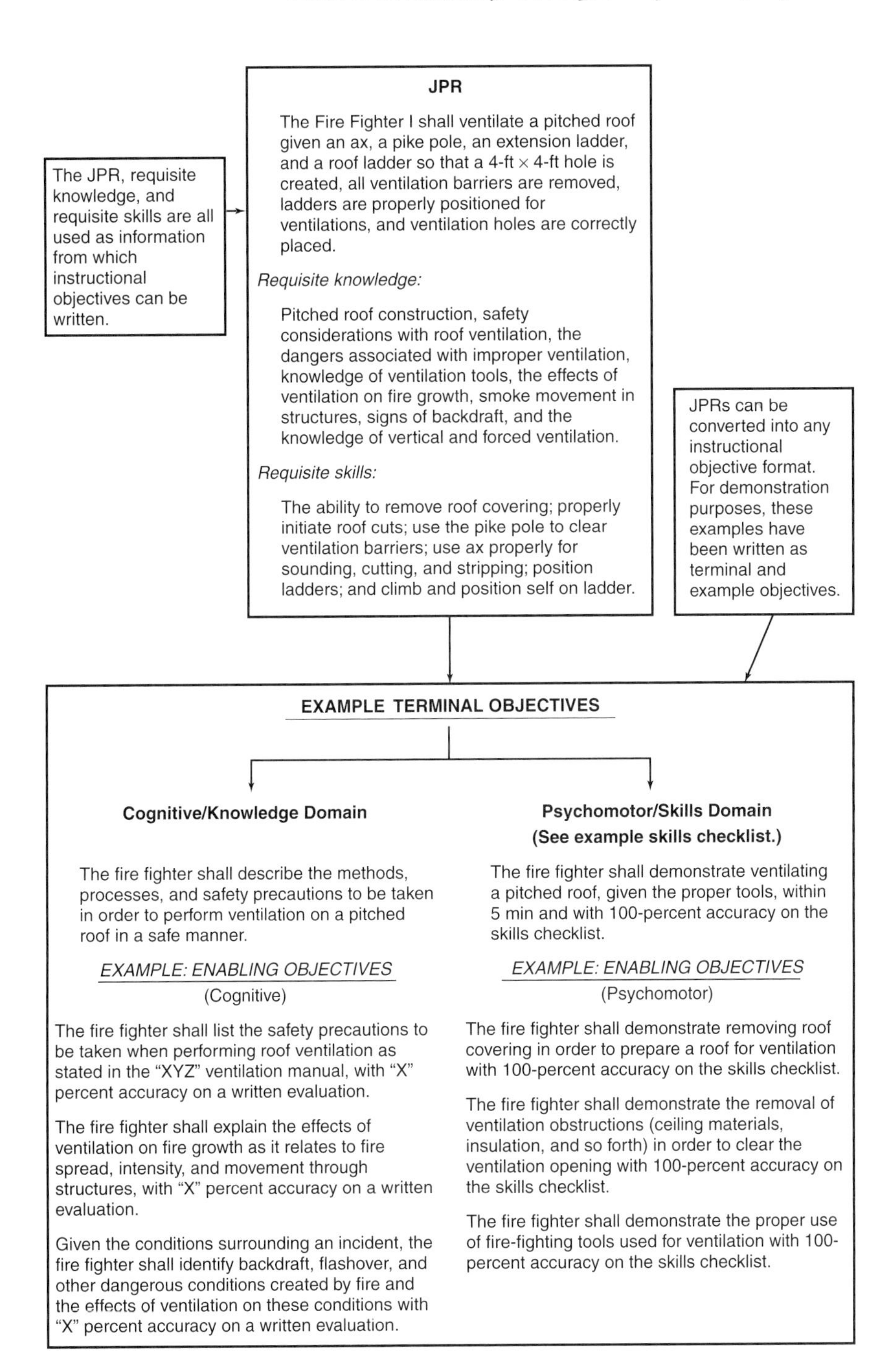

Figure B.3.2(a) Converting JPRs into Instructional Objectives.

Objective:
The fire fighter shall demonstrate ventilating a pitched roof, given the proper tools, within 5 min and with 100-percent accuracy on the skills checklist.

1. A 4-ft × 4-ft hole was created.	❑ Yes	❑ No
2. All ventilation barriers were removed.	❑ Yes	❑ No
3. Ladders were properly positioned.	❑ Yes	❑ No
4. Ventilation holes were correctly placed (directly over fire, at highest point, and so forth).	❑ Yes	❑ No
5. Task completed within 5 min. (Time to complete task: ________)	❑ Yes	❑ No

Figure B.3.2(b) Skills Checklist.

While the differences between job performance requirements and instructional objectives are subtle in appearance, the purpose of each statement differs greatly. JPRs state what is necessary to perform the job in the "real world." Instructional objectives, however, are used to identify what students must do at the end of a training session and are stated in behavioral terms that are measurable in the training environment.

By converting JPRs into instructional objectives, instructors will be able to clarify performance expectations and avoid confusion related to using statements designed for purposes other than teaching. Additionally, instructors will be able to add local/state/regional elements of performance into the standards as intended by the developers.

Requisite skills and knowledge should be converted into enabling objectives. These help to define the course content. The course content would include each of the requisite knowledge and skills. Using the above example, the enabling objectives would be pitched roof construction, safety considerations with roof ventilation, removal of roof covering, properly initiated roof cuts, and so on. This ensures that the course content supports the terminal objective.

Note that it is assumed that the reader is familiar with curriculum development or training design and evaluation.

B.4 Other Uses. While the professional qualifications standards are principally used to guide the development of training and certification programs, there are a number of other potential uses for the documents. Because the documents are written in JPR terms, they lend themselves well to any area of the profession where a level of performance or expertise must be determined. These areas might include the following:

(1) *Employee Evaluation/Performance Critiquing.* The JPRs can be used as a guide by both the supervisor and the employee during an evaluation. The JPRs for a specific job define tasks that are essential to perform on the job as well as the evaluation criteria to measure when those tasks are completed.
(2) *Establishing Hiring Criteria.* The professional qualifications standards can be used in a number of ways to further the establishment of hiring criteria. The AHJ could simply require certification at a specific job level—for example, Fire Fighter I. The JPRs could also be used as the basis for pre-employment screening by establishing essential minimal tasks and the related evaluation criteria. An added benefit is that individuals interested in employment can work towards the minimal hiring criteria at local colleges.
(3) *Employee Development.* The professional qualifications standards can be useful to both the employee and the employer in developing a plan for the individual's growth within the organization. The JPRs and the associated requisite knowledge and skills can be used as a guide to determine additional training and education required for the employee to master his or her job or profession.
(4) *Succession Planning.* Succession planning or career pathing addresses the efficient placement of people into jobs in response to current needs and anticipated future needs. A career development path can be established for targeted individuals to prepare them for growth within the organization. The JPRs and requisite knowledge and skills could then be used to develop an educational path to aid in the individual's advancement within the organization or profession.
(5) *Establishing Organizational Policies, Procedures, and Goals.* The JPRs can be incorporated into organizational policies, procedures, and goals where employee performance is addressed.

B.5 Bibliography.

Annett, John and Neville E. Stanton. 2001. *Task Analysis*. London and New York: Taylor and Francis.

Brannick, Michael T. and Edward L. Levine. 2001. *Job Analysis: Methods, Research and Applications for Human Resource Management in the New Millennium*. Conwin Press.

Dubois, David D., Ph.D. 1993. *Competency-Based Performance Improvement*. Amherst, MA: HRD Press.

Fine, Sidney A. and Steven F. Cronshaw. 1999. *Functional Job Analysis: A Foundation for Human Resources Management (Applied Psychology Series)*. Lawrence Erlbaum Association.

Gupta, Kavita. 1999. *A Practical Guide to Needs Assessment*. San Francisco, CA: Jossey-Bass/Pfeiffer.

Hartley, Darin E. 1999. *Job Analysis at the Speed of Reality*. Amherst, MA: HRD Press.

Hodell, Chuck. 2000. *ISD From the Ground Up*. Alexandria, VA: American Society for Training & Development.

Jonassen, David H., Martin Tessmer, and Wallace H. Hannum. 1999. *Task Analysis Methods for Instructional Design*. Lawrence Erlbaum Association.

McArdle, Gerie. 1998. *Conducting a Needs Analysis (Fifty-Minute Book)*. Crisp Publishing.

McCain, Donald V. 1999. *Creating Training Courses*. Alexandria, VA: American Society for Training & Development.

Phillips, Jack J. 2000. *In Action: Performance Analysis and Consulting*. Alexandria, VA: American Society for Training & Development.

Phillips, Jack J. and Elwood F. Holton III. 1995. *In Action: Conducting Needs Assessment*. Alexandria, VA: American Society for Training & Development.

Robinson, Dana Gaines and James C. Robinson. 1998. *Moving from Training to Performance: A Practical Guidebook*. San Francisco: Berrett-Koehler.

Schippmann, Jeffrey S. 1999. *Strategic Job Modeling: Working at the Core of Integrated Human Resources*. Lawrence Erlbaum Association.

Shepherd, Andrew. 2000. *Hierarchical Task Analysis*. London and New York: Taylor and Francis.

Zemke, Ron and Thomas Kramlinger. 1982. *Figuring Things Out: A Trainer's Guide to Task, Needs, and Organizational Analysis*. Perseus Press.

Annex C Comparison of NFPA 1001 1992 Edition Versus 1997 Edition

This annex, which is not a part of the requirements of NFPA 1001, is not included here.

Annex D Informational References

D.1 Referenced Publications. The following documents or portions thereof are referenced within this standard for informational purposes only and are thus not part of the requirements of this document unless also listed in Chapter 2.

D.1.1 NFPA Publications. National Fire Protection Association, 1 Batterymarch Park, P.O. Box 9101, Quincy, MA 02269-9101.

NFPA 1021, *Standard for Fire Officer Professional Qualifications,* 1997 edition.

NFPA 1061, *Standard for Professional Qualifications for Public Safety Telecommunicator,* 2002 edition.

NFPA 1403, *Standard on Live Fire Training Evolutions,* 2002 edition.

NFPA 1404, *Standard for Fire Service Respiratory Protection Training,* 2002 edition.

NFPA 1582, *Standard on Medical Requirements for Fire Fighters and Information for Fire Department Physicians,* 2000 edition.

NFPA 1962, *Standard for the Care, Use, and Service Testing of Fire Hose Including Couplings and Nozzles,* 1998 edition.

D.1.2 Other Publication.

Webster's Third New International Dictionary of the English Language, Unabridged. Springfield, MA: G. & C. Merriam Company, 1976.

D.2 Informational References. (Reserved)

D.3 References for Extracts.

The following documents are listed here to provide reference information, including title and edition, for extracts given throughout this standard as indicated by a reference in brackets [ ] following a section or paragraph. These documents are not a part of the requirements of this document unless also listed in Chapter 2 for other reasons.

NFPA 1500, *Standard on Fire Department Occupational Safety and Health Program,* 2002 edition.

Fire Service Organizations in the United States

The following associations are the most prominent groups serving the needs of the fire service in the United States.

Fire Department Safety Officers Association (FSSOA)
P.O. Box 149
Ashland, MA 01721-0149
www.fdsoa.org

The Fire Department Safety Officers Association is a not-for-profit association dedicated to promoting safety standards and practices in the fire and emergency services. Its purpose is to provide safety officers with the education necessary to perform their duties. Through these educational programs, FDSOA has taken steps toward decreasing the injury and death statistics associated with fire and emergency services.

Fire Marshals Association of North America (FMANA)
One Batterymarch Park
Quincy, MA 02269
www.nfpa.org

Organized in 1906, the Fire Marshals Association of North America was reorganized in 1927 as the Fire Marshals' Section of the National Fire Protection Association. The constitution and bylaws adopted

---

Source: From Appendix E, "Organizations with Fire Protection Interests in the United States" in A.E. Cote, ed., *Fire Protection Handbook*, 19th edition, NFPA, Quincy, MA, 2003.

March 31, 1947 (revised May 1982, May 1987, and May 1992), list the following objectives:

1. To unite for mutual benefit those officials engaged primarily in the prevention of fire, the control of arson, and/or the provision of public fire safety education
2. To provide educational and professional development opportunities through technological, certification, and prevention programs
3. To provide a resource service to the members of the Fire Marshals Association of North America
4. To actively market and promote a positive, dynamic, and proactive profile for the Fire Marshals Association of North America
5. To actively participate in the codes and standard-making process at the national, state, and local level
6. To expand the Fire Marshals Association of North America membership, create new chapters, and enhance membership participation at all levels
7. To monitor and support research and development of solutions to fire protection and fire prevention problems

Membership in the Fire Marshals Association of North America is contingent on membership in the NFPA and is open to the fire marshal, fire prevention officer, or such other official of each state, province, county, municipality, or fire protection district charged with the legal responsibility for fire prevention or investigation. Membership applications are available on request.

Member: The state, provincial, municipal, or local fire official who has been lawfully appointed and authorized by the authority having jurisdiction and charged with the statutory responsibilities and duties for fire prevention accomplished through enforcement of fire laws and regulations, property inspections, public fire safety education, or investigation of the cause and origin of fire.

Associate Member: Any individual who was previously an Association Member but cannot now qualify because of change in assignment or retirement.

Fire Service Section of the National Fire Protection Association

One Batterymarch Park
Quincy, MA 02269
www.nfpa.org

Organized in 1973, this organization is open to any member of the National Fire Protection Association who is any of the following:

1. An active or retired member of a fire department providing public fire prevention, fire suppression services, and/or EMS service to a state, county, municipality, or organized fire district
2. An active or retired member of a fire department providing fire prevention, fire suppression services, and/or EMS service to airfields and military bases
3. Principally engaged in the training and/or education of fire department members, whether or not an active or retired member of a particular department

Its objectives are as follows:

1. To unite for mutual professional benefit those members of NFPA who are members of the fire service
2. To act as a vehicle for the exchange of information among its members
3. To advance the interests of the profession in the fields of fire protection, public life safety education, emergency medical services, code enforcement, and fire suppression
4. To stimulate awareness of the need for continually improving programs in management, training, and education
5. To promote occupational health and safety issues by conducting meetings, conferences, seminars, and such other forums as may be practicable for the exchange of information and the encouragement of professionalism
6. To encourage participation of its members on the technical committees of the NFPA
7. To advance and encourage the development of improved fire suppression equipment, apparatus, and personal protective clothing ensembles
8. To encourage public authorities to specify and purchase fire protection equipment on the basis of performance standards
9. To bring to the attention of its members such matters of legislation and regulations as would be of interest
10. To promote cooperation within the fire service and between the fire service and other fire protection practitioners
11. To provide for the establishment, within the section, of a fire service professional society

International Association of Arson Investigators (IAAI)
12770 Boenker Rd.
Bridgeton, MO 63044
www.fire-investigators.org

The International Association of Arson Investigators (IAAI) was formed at Purdue University, West Lafayette, Indiana, in 1949, when insurance industry, fire service, and law enforcement personnel from the United States and Canada met to discuss the growing arson problem and the need for training and education in fire investigation. The association has the following objectives:

- To unite for mutual benefit those public officials and private persons engaged in the control of arson and kindred crime
- To provide for exchange of technical information and developments
- To cooperate with other law enforcement agencies and associations to further fire prevention and the suppression of crime
- To encourage high professional standards of conduct among arson investigators
- To continually strive to eliminate all factors that interfere with administration of crime suppression

Active membership in the IAAI is open to any representative (21 years of age or over) of government or of a governmental agency, and any representative of a business or industrial concern who is actively engaged in some phase of the suppression of arson and whose qualifications meet the requirements of the Membership Committee of the Association. Associate membership is open to persons not qualified for active membership after determination of their qualification by the Membership Committee.

The IAAI publishes a quarterly magazine, *The Fire and Arson Investigator*, and conducts an annual meeting in conjunction with a conference on fire investigation. Various committees are organized to assist the association in its attack on the arson problem, such as the Engineering Forensic Science, the Fraud Fire, the Insurance Advisory Committee, the Juvenile Firesetter Committee, the Training and Education Committee, and the Wildland Arson Committee.

International Association of Black Professional Fire Fighters (IABPFF)
8700 Central Ave., Ste. 306
Landover, MD 20785
www.iabpff.org

The IABPFF was organized in 1970 to (1) create a liaison between black fire fighters across the nation, (2) compile information concerning injustices that exist in the working conditions in the fire service and to implement action to correct them, (3) collect and evaluate data on all deleterious conditions where minorities exist, (4) see that competent blacks are recruited and employed as fire fighters where they reside, (5) promote interracial progress throughout the fire service, and (6) aid in motivating African Americans to seek advancement to elevated ranks.

International Association of Fire Chiefs (IAFC)
4025 Fair Ridge Dr.
Fairfax, VA 22033-2868
www.iafc.org

The International Association of Fire Chiefs (IAFC) is a professional organization open to all fire and emergency services administrators and managers, career and volunteer, interested in improving the delivery of fire and emergency medical services. The IAFC has over 12,000 members from all over the world, from rural volunteer departments to career metropolitan services. The association's mission is to provide vision, information, education, services, and representation to enhance the professionalism and capabilities of its members.

The IAFC is composed of eight geographical divisions as well as technical committees and sections that enable members to interact on a wide range of emergency services issues. The association currently has six sections: Apparatus Maintenance, EMS, Federal Military, Industrial and Fire Safety, Metropolitan Chiefs, and Volunteer Chief Officers. The IAFC is also served by over 14 committees.

Established in 1873, the IAFC has both a rich history and a strong commitment to the future. *IAFC On Scene* is a twice-monthly publication of the IAFC. Each year the IAFC sponsors Fire Rescue International, one of the largest fire and emergency services conferences/exhibitions in the world, in addition to several regional workshops.

International Association of Fire Fighters (IAFF)
1750 New York Ave., NW
Washington, DC 20006-5395
www.iaff.org

Organized in 1918, the IAFF has approximately 240,000 members in the United States and Canada. The IAFF is affiliated with the American Federation of Labor and Congress of Industrial Organizations in the

United States and the Canadian Labour Congress (AFL-CIO/CLC). Any person, who is engaged as a permanent and paid employee of a fire department, is eligible for active membership through the chartered locals, state or provincial associations, and joint councils. Conventions of the Association are biennial. The IAFF provides a variety of services to the membership, including technical assistance in the occupational safety and health area, labor relations expertise through in-house personnel and field representatives, an extensive educational seminar program, legislative representation at the national level, and public relations guidance.

International Fire Service Accreditation Congress (IFSAC)
Oklahoma State University
1700 W. Tyler
Stillwater, OK 74078-8075
www.ifsac.org

The International Fire Service Accreditation Congress is a nonprofit, peer-driven, self-governing system of both fire service certification programs and higher education fire-related degree programs. IFSAC, which was founded in 1990 to ensure the quality and continuation of a national accreditation system for fire fighter certification programs, developed an accreditation system for fire-related degree programs in 1992. IFSAC is represented in 5 countries and in 55 states and provinces and has as its mission to plan and administer a high-quality, uniformly delivered accreditation system with an international scope.

Within a congress of the whole, which deals with the general business of IFSAC, two separate assemblies specialize in the issues peculiar to degree-granting institutions and certifying entities. The IFSAC Certificate Assembly provides accreditation to entities that certify the competency of and issue certificates to individuals who pass examinations based on National Fire Protection Association fire service professional qualifications and other standards approved by the assembly. The accreditation is made at the state, provincial, federal government, or territorial level for fire fighter certification programs. The IFSAC Degree Assembly accredits fire science or related academic programs at colleges and universities. Accreditation includes both two-year associate degrees and four-year bachelor's degrees.

Each assembly has a Board of Governors, elected from the membership, which acts on all accreditation applications. In both assemblies, accreditation is granted only after an integral self-study is conducted by the entity or institution seeking accreditation, followed by an on-site review by a panel consisting of peer representatives from other member entities or institutions. Administrative offices located at Oklahoma State University handle the daily operations of the organization.

International Fire Service Training Association (IFSTA)
Fire Protection Publications
Oklahoma State University
930 N. Willis St.
Stillwater, OK 74078-8045
www.ifsta.org

The International Fire Service Training Association was established in 1934 as a nonprofit educational association of fire-fighting personnel who are dedicated to providing accurate and up-to-date training material for the fire service. IFSTA's purpose is to validate training materials for publication by checking drafts for errors, adding information on new techniques and developments, and deleting obsolete or dangerous information. To carry out the IFSTA mission, Fire Protection Publications was established as an entity of the College of Engineering, Architecture, and Technology at Oklahoma State University. Fire Protection Publications' primary function is to publish and distribute the IFSTA-validated training manuals. Fire Protection Publications also produces and/or markets a wide variety of other books, curricula, audiovisual materials, and computer software for the fire service training arena. The IFSTA manuals have been adopted as the official training texts in most of the states and provinces of North America and by numerous foreign countries. Foreign language versions of some of the more popular IFSTA manuals are available.

International Municipal Signal Association (IMSA)
P. O. Box 539
Newark, NY 14513
www.imsasafety.org

IMSA, a leading international resource for information, education, and certification in public safety, was organized in 1896 and currently has over 10,000 members. It is an educational, nonprofit organization dedicated to conveying knowledge, technical information, and guidance to its membership, which consists of municipal signal, signs, and communication department heads and their employees.

The range of communications covered includes traffic control, fire alarm, and public safety dispatchers. There are twenty-two sections of IMSA (based on geographical areas in the United States and Canada), as well as a Sustaining Section. The former serve the regional needs of its members. The *IMSA Journal* is the association's bimonthly publication. Certification programs are offered for both interior and municipal fire alarms.

International Society of Fire Service Instructors (ISFSI)
P.O. Box 2320
Stafford, VA 22555
www.isfsi.org

The ISFSI, organized in 1960, is composed of persons responsible for the training of fire officers, fire fighters, and rescue and emergency medical personnel. The society's goal is to assist in the development of fire service instructors through better training and educational opportunities, to provide the means for continuous upgrading of instructors through in-service training, and to actively promote the role of the fire service instructor in the total fire service organization in industry and the public sector. It has membership in all fifty states and fifteen foreign countries.

Metropolitan Fire Chiefs Section of the National Fire Protection Association (NFPA) and the International Association of Fire Chiefs (IAFC)
www.nfpa.org/metro

The executive secretary of the NFPA Section can be reached at 3257 Beals Branch Rd., Louisville, KY 40206. The IAFC Section can be reached at 4025 Fair Ridge Dr., Suite 300, Fairfax, VA 22033-2868. Membership in the Metropolitan Fire Chiefs Section is limited to members of the NFPA and the IAFC who are fire chiefs of cities or jurisdictions having a minimum authorized strength of 400 fully paid, uniformed personnel. Membership classifications include regular (active) members, senior (retired) members, honorary members, and affiliate members. The purpose of the Metro Chiefs, originally organized in 1965, is to bring together fire chiefs from large, metropolitan fire departments in order to share information and to affect policy changes.

National Board on Fire Service Professional Qualifications (Pro Board)
P.O. Box 690632
Quincy, MA 02269
www.npqs.win.net

The mission of the National Board on Fire Service Professional Qualifications (Pro Board) is to establish an internationally recognized means of acknowledging professional achievement in the fire services and related fields. The certification of uniformed members of public fire departments, both career and volunteer, is the primary goal. However, other individuals and organizations with fire protection interest may also be considered for participation. The Pro Board accredits state, provincial, educational, and governmental institutions that test and certify emergency services personnel to the National Fire Protection Association's fire service professional qualifications standards. The Pro Board also maintains a National Registry of those certified to these standards.

National Volunteer Fire Council (NVFC)
1050 17th St., NW, Ste. 490
Washington, DC 20036
www.nvfc.org

The NVFC is a nonprofit membership association representing the interests of the volunteer fire, EMS, and rescue services. Organized in 1976, the NVFC serves as the information source regarding legislation, standards, and regulatory issues. Its membership includes state-level organizations that represent volunteer fire fighters and EMS personnel in 49 states, individual fire fighters, fire departments, and corporate members.

Its purpose is to represent the volunteer fire and emergency medical services in the national policy arena and on numerous national and international committees and organizations. The NVFC also serves on the Board of Visitors of the National Fire Academy, the Fallen Firefighters Foundation, the Congressional Fire Services Institute Advisory Board, and the Federation of World Volunteer Firefighters Association.

Women in the Fire Service, Inc. (WFS)
P.O. Box 5446
Madison, WI 53705
www.wfsi.org

WFS is a nonprofit organization providing networking, advocacy, and peer support for women in the fire service and information resources on women's issues to the fire service at large. Established in 1982, WFS offers publications (both brochures and periodicals), workshops, consulting, and national conferences on topics relating to the gender integration of the fire service.